Make: JavaScript Robotics

Rick Waldron, Anna Gerber, David Resseguie, Emily Rose, Susan Hinton, Sara Gorecki, Bryan Hughes, Andrew Fisher, Julian David Duque, Pawel Szymczykowski, Donovan Buck, Jonathan Beri, Kassandra Perch, Raquel Vélez, Lyza Danger Gardner

Make: JavaScript Robotics

by Backstop Media and Rick Waldron

Printed in Canada.

Published by Maker Media, Inc., 1160 Battery Street East, Suite 125, San Francisco, California 94111.

Maker Media books may be purchased for educational, business, or sales promotional use. Online editions are also available for most titles (*http://my.safaribooksonline.com*). For more information, contact our distributor's corporate/institutional sales department: 800-998-9938 or *corporate@oreilly.com*.

Editor: Brian Jepson
Production Editor: Melanie Yarbrough
Copyeditor: Tracy Brown Hamilton
Proofreader: Jasmine Kwityn
Indexer: Meghan Jones, WordCo Indexing
Interior Designer: David Futato
Cover Designer: Brian Jepson
Cover Photographer: Pawel Szymczykowski
Illustrator: Rebecca Demarest

April 2015: First Edition

Revision History for the First Edition

2015-04-03: First Release

See *http://oreilly.com/catalog/errata.csp?isbn=9781457186950* for release details.

978-1-4571-8695-0

[TI]

Table of Contents

Preface

By Rick Waldron

I love programming and I also love making things. When I discovered Chris Williams' `node-serialport` module (for Node.js), I remember thinking, "Now I can program the things that I make." So I did! I contributed an Arduino sketch that printed a photoresistor sensor value to the open serial port and a small JavaScript handler program that listened for data and printed a "graph" to the terminal. Months later, I started contributing to Cam Pedersen's duino project, until Julian Gaultier approached me with his JavaScript implementation of the Firmata protocol. From there, we set out to build a collection of high-level component classes with one goal: to make it easy and fun to control hardware with JavaScript. This book will show you what we built and how to use it to program the things that *you* make.

While the physical challenges of engineering a hardware project remain the same as they would for a project programmed in any other language, this book is going to show you how to think about that hardware in terms of objects that maintain state and provide control behaviors in the form of intuitively designed interfaces.

So, how would you verbally describe adding an LED to a project and then turning it on? You might say, "Connect the LED to ground and pin 9, then turn it on." Using the Johnny-Five framework, that would be written as:

```
// Connect the LED to ground and pin 9
var led = new five.Led(9);

// then turn it on
led.on();
```

What about connecting a servo and then setting its horn to a specific angle in degrees? "Attach the servo to pin 10 and position its horn to 110°." Here's what that looks like:

```
// Attach the servo to pin 10
var servo = new five.Servo(10);

// position its horn to 110°
servo.to(110);
```

These examples both illustrate an output, but what about input? Consider how Arduino sketches work: they generally rely on a *program loop* and often introduce some form of delay when reading and processing input. This means that your Arduino sketch is *blocked* whenever it is waiting for input. When writing programs in JavaScript, the process is *never* blocked; instead, your handlers wait for data to arrive and process it asynchronously:

```
var sensor = new five.Sensor("A0");

sensor.on("data", function() {
  console.log(this.value);
});
```

These are trivial examples, but they illustrate the patterns that you will see repeated consistently throughout this book. Each project will show you how to construct it in the physical sense, then construct it in the abstract programming sense, and the latter will align with the former.

With these concepts, you will build and program:

- Walkers, typers, swimmers, and rovers (Chapters 1, 2, 3, and 4)
- A dancing hexapod (Chapter 5)
- Voice-activated relay control (Chapter 6)
- An indoor sundial (Chapter 7)
- Holiday, mood, or anytime lighting (Chapters 8 through 10)
- A security and notification system (Chapter 11)
- Sonar-based artificial intelligence (Chapter 12)
- A delta bot (Chapter 13)
- Musical shoes (Chapter 14)

For me, the most exciting part about this book is the authors themselves. This group is an excellent representation of NodeBots community members that have stood out since the very beginning. They are more than just writers or engineers: they are teachers, communicators, leaders, and (in my opinion) heroes. It would be an understatement to say that I couldn't have done it without them.

Enough talk, more rock. These projects do not have to be done in any specific order, so take a look at the Table of Contents, find a project that sounds like fun, and start building!

Conventions Used in This Book

The following typographical conventions are used in this book:

Italic
: Indicates new terms, URLs, email addresses, filenames, and file extensions.

`Constant width`
: Used for program listings, as well as within paragraphs to refer to program elements such as variable or function names, databases, data types, environment variables, statements, and keywords.

`Constant width bold`
: Shows commands or other text that should be typed literally by the user.

`Constant width italic`
: Shows text that should be replaced with user-supplied values or by values determined by context.

This element signifies a tip, suggestion, or general note.

This element indicates a warning or caution.

The part numbers in each chapter use the following abbreviations:

- MS: Maker Shed (*http://makershed.com*)
- AZ: Amazon (*http://amazon.com*)
- AF: Adafruit (*http://adafruit.com*)
- SF: SparkFun (*http://sparkfun.com*)

Using Code Examples

This book is here to help you get your job done. In general, you may use the code in this book in your programs and documentation. You do not need to contact us for permission unless you're reproducing a significant portion of the code. For example, writing a program that uses several chunks of code from this book does not require permission. Selling or distributing a CD-ROM of examples from Make: books does require permission. Answering a question by citing this book and quoting example code does not require permission. Incorporating a significant amount of example code from this book into your product's documentation does require permission.

We appreciate, but do not require, attribution. An attribution usually includes the title, author, publisher, and ISBN. For example: "*Make: JavaScript Robotics* by Jonathan Beri, Donovan Buck, Julian David Duque, Andrew Fisher, Lyza Danger Gardner, Anna Gerber, Sara Gorecki, Susan Hinton, Bryan Hughes, Kassandra Perch, David Resseguie, Emily Rose, Pawl Szymczykowski, Raquel Velez, Rick Waldron (Maker Media). Copyright 2015 Backstop Media, 978-1-4571-86950."

If you feel your use of code examples falls outside fair use or the permission given here, feel free to contact us at bookpermissions@makermedia.com.

Safari® Books Online

Safari Books Online is an on-demand digital library that delivers expert content in both book and video form from the world's leading authors in technology and business.

Technology professionals, software developers, web designers, and business and creative professionals use Safari Books Online as their primary resource for research, problem solving, learning, and certification training.

Safari Books Online offers a range of plans and pricing for enterprise, government, education, and individuals.

Members have access to thousands of books, training videos, and prepublication manuscripts in one fully searchable database from publishers like O'Reilly Media, Prentice Hall Professional, Addison-Wesley Professional, Microsoft Press, Sams, Que, Peachpit Press, Focal Press, Cisco Press, John Wiley & Sons, Syngress, Morgan Kaufmann, IBM Redbooks, Packt, Adobe Press, FT Press, Apress, Manning, New Riders, McGraw-Hill, Jones & Bartlett, Course Technology, and hundreds more. For more information about Safari Books Online, please visit us online.

How to Contact Us

Please address comments and questions concerning this book to the publisher:

Maker Media, Inc.
1160 Battery Street East, Suite 125
San Francisco, CA 94111

Make: unites, inspires, informs, and entertains a growing community of resourceful people who undertake amazing projects in their backyards, basements, and garages. Make: celebrates your right to tweak, hack, and bend any technology to your will. The Make: audience continues to be a growing culture and community that believes in bettering ourselves, our environment, our educational system—our entire world. This is much more than an audience; it's a worldwide movement that Make: is leading—we call it the Maker Movement.

For more information about Make:, visit us online:

- Make: magazine: *http://makezine.com/magazine*
- Maker Faire: *http://makerfaire.com*
- Makezine.com: *http://makezine.com*
- Maker Shed: *http://makershed.com*

All source code for the examples in this book can be found at: *https://github.com/rwaldron/javascript-robotics*.

We have a web page for this book, where we list errata, examples, and any additional information. You can access this page at: *http://bit.ly/1KUV1p2*.

If you feel your use of code examples falls outside fair use or the permission given above, feel free to contact us at *permissions@oreilly.com*.

Acknowledgments

Julián Duque

NodeBots is all about learning while having fun and especially for me it's a way to achieve social impact. Everything started at NodeConf 2013—it was my first time using Johnny-Five on a NodeBots workshop guided by Rick Waldron and Raquel Velez. My first attempt was with a piezo buzzer, but at that time the library didn't support the `Piezo` class, so I asked Rick how to use it and he said to me, "I wasn't able to make it work, how about a pull request?" That's how I started contributing to the project. On the other hand, Raquel gave me my first Arduino kit as a gift and told me, "Take this with you and teach what you've learned in Colombia." That humble action was so inspiring for me and I started a NodeBots chapter in Medellín, Colombia. I also helped create the communities in Uruguay, Guatemala, and Mexico through workshops and talks. Rick and Raquel, thank you so much for inspiring me and letting me be part of this Maker revolution.

Building Robots with Lo-tech Materials

1

By Andrew Fisher

When you think about robots, you probably imagine drones, self-driving cars, or humanoid robots like Atlas or Asimo. Many of these more serious robots start their costs at thousands of dollars, and there is no real upper limit (Atlas costs over a $1M, for example). It is possible, however, to build small, interesting robots with a few inexpensive electronic components coupled with some materials you can readily find at home.

One of the great things about NodeBots is being able to prototype rapidly. The combination of an approachable language like JavaScript with friendly hardware such as Arduino means you can explore ideas and see how things work. Being able to prototype and play with robotic techniques quickly helps with learning and exploring concepts.

In this chapter, we'll explore several robotic concepts using a basic robot called the SimpleBot (Figure 1-1). The first design of the SimpleBot came as the result of a challenge from my child, who asked to build a robot together one evening after dinner. With the clock ticking and only an hour to work before bedtime, it meant using things we had on hand—no laser cutters, CNC mills, or 3D printers. In true hacker spirit, we fabricated using cardboard, cable ties, and rubber bands to get something that worked.

Figure 1-1 *Completed SimpleBot*

Figure 1-2 shows the SimpleBot we made that night. After building it, we fell in love with prototyping using materials such as cardboard, cable ties, and more recently, corflute (corrugated plastic board) for robotics. These materials are inexpensive and easy to work with using scissors or a good craft knife. You don't have to have access to tools such as a laser cutter, though if you do, then you can still work with these materials—it just becomes even faster to cut things out (and a bit more accurate). Working with these materials allows you to fabricate on your kitchen table and kids can easily work with them, too.

Figure 1-2 *The very first SimpleBot*

After that first effort, the SimpleBot has gone through numerous revisions and is now used as a teaching robot for some NodeBot events. I hope I've convinced you that building robots out of simple materials such as cardboard is a good idea. This chapter is going to cover:

- Building the basic SimpleBot platform
- Cutting the cord and untethering our SimpleBot from our computer

Building the SimpleBot

Before you get building, remember there's no right or wrong way to build your SimpleBot. The design of the SimpleBot was intentionally left open-ended so you can make it any way you want. Others have built versions as minimal as possible on one extreme, as well as automated Nerf-Gun-Toting platforms on the other. The point of the SimpleBot is to play, explore, and extend it, to further your understanding of robotics—so customize away.

Bill of Materials

The SimpleBot project is divided into two parts with components needed at each stage. All of the components for this chapter are listed in Table 1-1, and then the elements needed for each stage are listed again when you get to that point in the chapter. Table 1-2 lists the parts needed for the wireless version.

Table 1-1 *SimpleBot materials*

Count	Part	Estimated price	Part numbers/source
1	Arduino Uno - R3	$24.95	MS MKSP99; AF 50; SF DEV-11021
1	Half-sized breadboard	$5	MS MKKN2, SF PRT-12002, AF 64
Multi	Jumper wires (male to male)	$4.95	MS MKSEEED3, SF PRT-11026, AF 758
Multi	Jumper wires (male to female)	$5	MS MKKN5, SF PRT-09385, AF 825
2	Continuous rotation (CR) servos; if you have standard servos, an Adafruit tutorial (*http://*	$14	MS MKPX18, AF 154, SF ROB-09347

Count	Part	Estimated price	Part numbers/source
	bit.ly/19LXNRw) shows how to mod them into CR servos		
15	Cable ties (3–4mm wide and about 200mm long is perfect)	$1	Various suppliers/stores
1	7.4v RC battery pack (LiPo, Li-Ion, or whatever you can get a hold of)	$5	Old or broken RC toys are good sources of these
1	LM7806 or NTE962 6V voltage regulator (drops the voltage to something the servos can use)	$2	RadioShack or Amazon
1	Chassis material—a square of thick cardboard or corflute (3–5mm thick), approximately 400mm square	$2	Various suppliers/stores
1	A printout of the template file	Free	Included with source code

Table 1-2 *Wireless SimpleBot materials*

Count	Part	Notes	Estimated price
4x	50V 0.1uF ceramic capacitors	Ceramic is best, but others will work	$0.50
1x	USR WiFi232-T module	*http://www.usr.so*	$15

Build Steps

1. Start by cutting out the template on your cardboard. It doesn't need to be perfect. A knife, a cutting mat, and a ruler make it easier to do the inside holes than scissors. With the wheels, the center hole can be cut out or just left, it's only a target so you know where the center is so you can screw through it to the servo.

   You should now have a full set of chassis pieces. The large piece with the bumps at the end is the base, and the bumps are the front. Use the bumps to push through the small holes on the smaller rectangle, which you can use as a bumper to mount things on.

2. Next, mount the wheels to the servo. Take the cross-shaped servo horn (you should have a packet of different shaped plastic fittings with your servo) and align the center with the center of the wheel. You can use a piece of wire or a needle to poke small holes in your material to mark the points to screw (Figure 1-3).

   Screw the wheels securely to the servo horn. If you prefer, just glue them on (but then you can't reuse the servo horns later)—either way works fine. Once you have the wheels mounted on the servo horns, screw through the center of the wheel and the horn into the small gear on the servo. Go easy and hold everything in place while you turn the screwdriver; being rough here can strip the gears in your servo and cause them to slip. Do this for both servos so you now have two wheels.

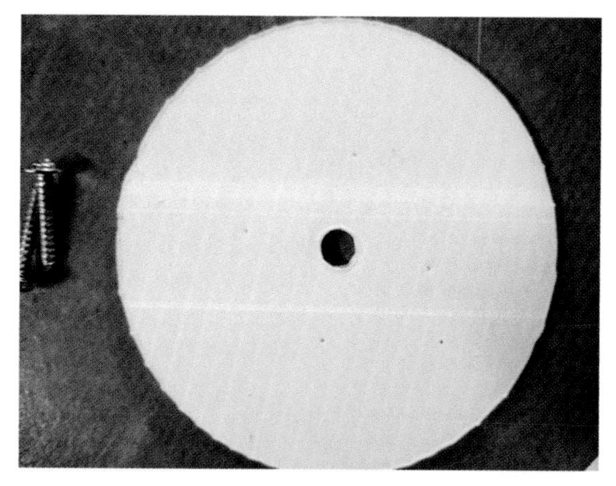

Figure 1-3 *Mark screw points using the servo horn and a piece of wire*

Gently rotate the servo to ensure it turns freely. If you're using cardboard or corflute, your mounting screws may be a little long. They don't have to go all the way through, but make sure they clear the body of the servo when you turn them.

3. Next, mount the servos to the chassis, as shown in Figure 1-4. You want to place these more or less to the front so the weight balances. You can put them anywhere, but bear in mind you may need to weigh down one end if you find it's tipping. Mount the servos so the side with the wheel is closest to the front of the chassis, and attach them with two cable ties through the mounting holes to keep them in place. Use two cable ties so the servo body doesn't twist when you start driving.

When you attach the cable ties, only tighten them enough to stop movement, but not so tight as to rip the cardboard or corflute.

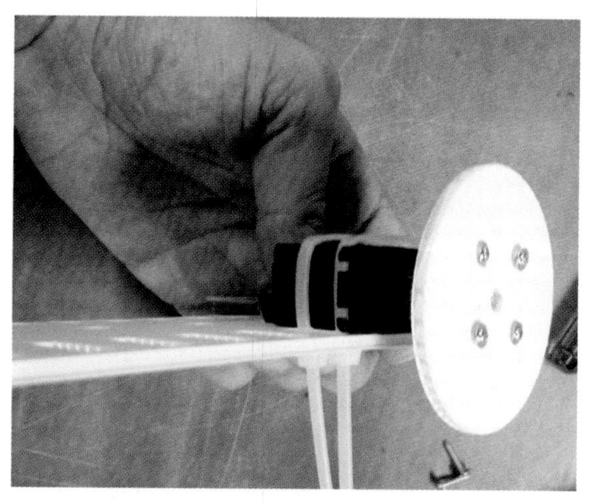

Figure 1-4 *Servo mounted to the chassis*

4. Mount the battery between the wheels as in Figure 1-5. Again, you can use a cable tie, but double-sided tape or a bit of Blu-Tack works well here, too, if you want to recharge easily. Trim the cable tie excess off as you go or it gets a little hard to mount everything.
5. Next, fashion a simple "skid" for the SimpleBot to stay balanced. You can do this by looping a cable tie toward the back of the SimpleBot in the middle of the chassis. Do this from the top so the catch of the tie doesn't get caught on anything. The underside of the loop should be about the same height as half the wheel so the body sits level. It will feel "loose," but don't worry, you secure this with the breadboard in the next step.

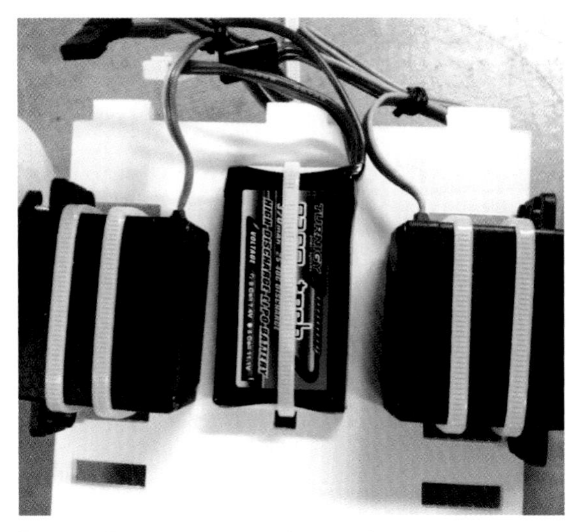

Figure 1-5 *Battery mounted to the chassis*

6. Finally, add the breadboard. Mount this across the point you put the skid cable tie so the board holds the skid down. The cable tie can go down the length of the board where the channel is and you can still place ICs across it.

   Now that you've finished the mechanics, it's time for the electronics. The complete wiring diagram is shown in Figure 1-6, with each piece explained next.

7. Start by creating a battery power rail. This goes to the back of the breadboard and gives you at least 7.4V. Join the grounds on both sides of the breadboard together. Mount the Arduino Nano at one end and join its ground. As you can see with the Nano, position it to either side of the main channel, over the cable tie, and position the USB connection on one side so it's easy to plug in, as shown in Figure 1-7.

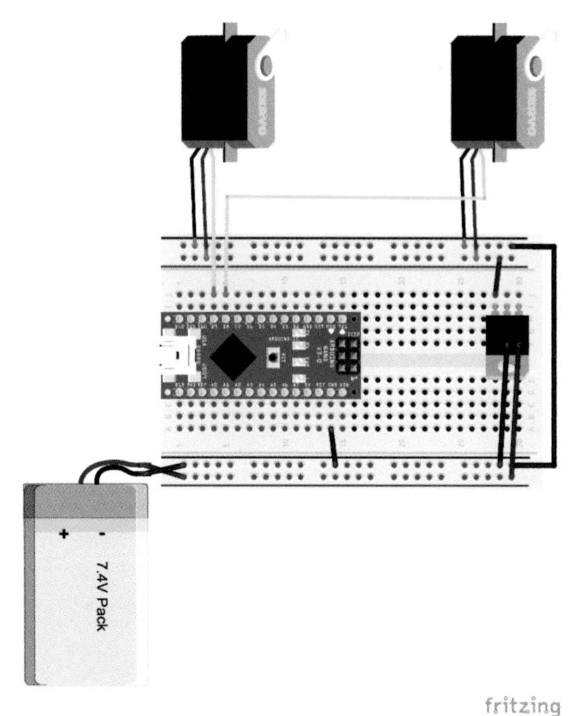

Figure 1-6 *SimpleBot wiring diagram*

Figure 1-7 *Distributing power*

8. Now create the power for the servos. The servos want 6V so the battery will give them a bit too much, so use a +6V voltage regulator to output a nice clean 6V for the servos. Put that 6V on the other rail of the breadboard so you can attach the servos there.

Note that we're using the breadboard to bring power from the battery to the regulator from "behind," then power from the regulator to the other rail from the "front." This is a good way of keeping what is going on straight in your head when it comes time to debug the circuit later in this chapter.

9. The servos can be attached to the front power rail with voltage and ground going to each. The left servo signal wire goes to pin 9 and the right servo signal wire goes to pin 8 on the Arduino, as shown in Figure 1-8.

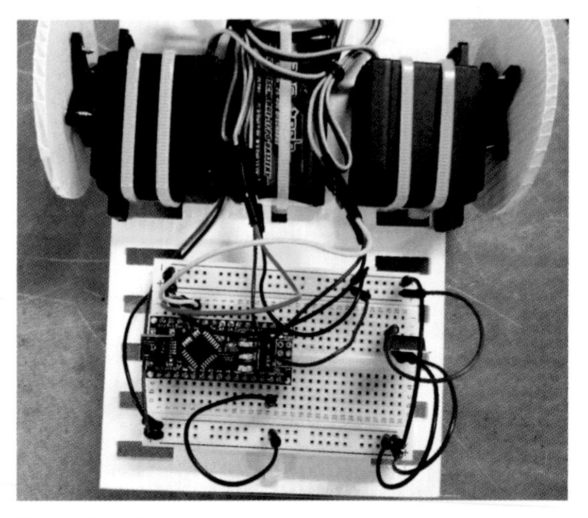

Figure 1-8 *Servo wiring*

Give yourself a pat on the back and tidy up all the wires so nothing is dangling on the floor—you are now ready to start working on your code.

All source code for the examples in this book can be found on GitHub (*http://bit.ly/19LX9n3*).

Installing Node.js Packages

Ensure Node.js is installed (see "Installing Node.js"). You only need a couple of packages for the SimpleBot, which can be installed from a terminal shell with:

```
npm install johnny-five temporal keypress
```

Testing the Build with a Basic Program

Your Arduino needs Firmata on it (see "Arduino") so the first thing to do is test that everything works before building something more complex to control the SimpleBot. The following code connects to the Arduino over USB and then runs the servos forward for 3 seconds, stops for 3 seconds, goes backward for 3 seconds, and finally stops before exiting.

Connect your SimpleBot to your computer and run the script shown in Example 1-1 and replace `<serialport>` with the serial port your Arduino is connected to.

If this all works, great! Your SimpleBot is working properly.

Troubleshooting

If your SimpleBot isn't working as described, don't worry, it is time for some debugging:

Neither of the servos turn
: Check your wiring.

  Each servo should be connected to the power rail from the battery and to ground on the red and brown (or black) wires. The left servo's signal wire should be connected to pin 8 on the Arduino and the right servo's signal wire connected to pin 9 on the Arduino.

  Also ensure that the battery's ground is connected to the Arduino's ground, as per the wiring diagram—you need a common ground across power sources.

One of the servos doesn't turn
: Check your wiring with a particular focus on ensuring your servo signal line is wired into the correct pin on the Arduino.

Example 1-1 *servo-test.js (servo testing code)*

```
var five = require("johnny-five");
var temporal = require("temporal");

var opts = {};
opts.port = process.argv[2] || "";

var board = new five.Board(opts);

board.on("ready", function() {

  var left_wheel = new five.Servo.Continuous(9);
  var right_wheel = new five.Servo.Continuous(8);

  temporal.queue([
    {
      delay: 5000,
      task: function() {
        console.log("going forward");
        left_wheel.cw();
        right_wheel.ccw();
      }
    }, {
      delay: 3000,
      task: function() {
        console.log("stopping");
        left_wheel.stop();
        right_wheel.stop();
      }
    }, {
      delay: 3000,
      task: function() {
        console.log("going backward");
        left_wheel.ccw();
        right_wheel.cw();
      }
    }, {
      delay: 3000,
      task: function() {
        console.log("stopping");
        left_wheel.stop();
        right_wheel.stop();
      }
    }, {
      delay: 1500,
      task: function() {
        console.log("Test complete. Exiting.");
        process.exit();
      }
    }
  ]);
});
```

The SimpleBot spins on the spot
: This is caused by a servo rotating in a different direction than is assumed in the code. You can unhook the servo and flip it over, which will make all the other code samples behave. Otherwise, just modify the code to change the offending servo's command from `servo.ccw()` to `servo.cw()` and vice versa.

A Simple Driving Program

Now that the SimpleBot is working correctly, you can create a simple program to drive it around. As a starting point, let's take input from the keyboard in order to drive the SimpleBot forward and backward, spin it left or right, and also to stop it. The code shown in Example 1-2 is a very basic example of how to drive using keyboard control.

Example 1-2 *simplebot.js (driving example)*

```
var five = require("johnny-five");
var keypress = require("keypress");
keypress(process.stdin);

var opts = {};
opts.port = process.argv[2] || ""; ❶
var board = new five.Board(opts);

board.on("ready", function() {

  console.log("Control the bot with the arrow keys, the space bar to stop, Q to exit.")

  var left_wheel = new five.Servo.Continuous(9);
  var right_wheel = new five.Servo.Continuous(8);

  // Configure stdin for the keyboard controller
  process.stdin.resume();
  process.stdin.setEncoding("utf8");
  process.stdin.setRawMode(true);

  process.stdin.on("keypress", function(ch, key) { ❷

    if (!key) {
      return;
    }

    if (key.name == "q") {
      console.log("Quitting");
      process.exit();
    } else if (key.name == "up") { ❸

      console.log("Forward");
      left_wheel.cw();
      right_wheel.ccw();

    } else if (key.name == "down") {

      console.log("Backward");
```

```
            left_wheel.ccw();
            right_wheel.cw();

        } else if (key.name == "left") {

            console.log("Left");
            left_wheel.ccw();
            right_wheel.ccw();

        } else if (key.name == "right") {

            console.log("Right");
            left_wheel.cw();
            right_wheel.cw();

        } else if (key.name == "space") {

            console.log("Stopping");
            left_wheel.to(90);
            right_wheel.to(90);
        }
    });
});
```

❶ Connect to the board using the serial connection supplied from the command line.

❷ Set up an event waiting for a keypress on the keyboard.

❸ Check the key that was pressed and then, depending on which direction you want to go, engage the motors as discussed in the section on differential drive.

Connect your SimpleBot to your computer and run the script with:

```
node simplebot.js &lt;serialport&gt;
```

You should now be able to drive your SimpleBot around the table or floor using the arrow keys on your keyboard and hitting the spacebar to stop.

Troubleshooting

Here are some tips in case you get stuck:

One or both of the servos don't move

See the previous troubleshooting section on tracking down wiring issues.

You hit the space bar to stop, but the robot keeps moving a little bit

This is due to continuous rotation servos being a hack on normal servos, as well as each servo being manufactured slightly differently. Some servos have a "tuning pot" at the back, which you can turn so that it stops when you set it to stop.

If you don't have one of these, then you can set your stop point in code.

On a CR servo, the stop position is defined as the center (recall a normal servo operates in an arc that is usually 180°), which is usually 90°. Setting a lower value than this will rotate the servo one direction and above this will rotate the other.

In the stop code, set the servo to move to the 90° position. As such, you might have to tune your stop point a little, changing the stop code to be like this:

```
left_servo.to(87);
right_servo.to(94);
```

There's no "correct" number and every servo will be slightly different, so a few minutes of tinkering will enable you to figure out the number you need for that servo.

Drive Considerations

The SimpleBot drive system is based on a method called *differential drive*, where each wheel is controlled independently by the servo motor that is attached to it. This is a very common design for two-wheeled robots, because the two wheels control both forward and backward motion as well as overall direction. In its most basic form, this makes the robot extremely simple to control.

- To go forward, we rotate both wheels in the forward direction.
- To go backward, we rotate both wheels in the reverse direction.
- To turn left, we stop or rotate the left wheel backward and drive the right wheel forward.
- To turn right, we stop or rotate the right wheel backward and drive the left wheel forward.

The biggest implication of this design is that your robot typically spins on the spot when you want to change direction. Turning in an arc, rather than on the spot, is possible (by slowing down one wheel relative to the other), but results in considerably more code to make it work.

Contrast this differential drive design to the way a car drives: two wheels (or four, if the vehicle has all-wheel drive) are rotated by a single motor via an axle to create drive and then two wheels are oriented at different angles in order to change direction.

Different drive systems have advantages and disadvantages, but differential drive is a good introduction.

Cutting the Cord

Now that you have your SimpleBot driving around the table, you'll notice one big limitation—the length of the USB cable tethering your bot to your computer. Obviously one solution is to use a longer cable; however, anything over 3 meters long starts having signal issues, not to mention the potential of tangles and the weight the robot has to drag around.

A better solution is to cut the cord altogether—letting the SimpleBot roam free and unshackled from its USB tether!

Building a Wireless SimpleBot

For this build, we'll use a single battery with bypass filters to smooth the voltage spikes from the servos as a result of everything using just the one battery. We'll also use WiFi, as this will give a little more range than Bluetooth and allows driving around the workspace or home or wherever the WiFi network exists.

The WiFi module used in this circuit is very inexpensive at about $12 and is quite a clever bit of circuitry. The module operates as a WiFi-to-Serial bridge so anything sent to it over the network gets copied to its serial lines and vice versa. This means our Arduino setup is almost exactly the same as before, but it can go wireless with the addition of this module—a great result for $12, and one change to the code.

Bill of Materials

The specific items you'll need for this stage are listed here:

- 50V 0.1uF ceramic capacitors
- M-M jumper wires
- M-F jumper wires
- USR WiFi232-T module

Wiring Up

Before starting, unplug your Arduino from the USB so you don't short anything out while building this circuit.

What sort of wireless?

In this project, we've gone for WiFi, but there are many other types of radio you can choose, such as Bluetooth and 433MHz serial. Each radio type has its own strengths, weaknesses, and trade-offs. The key things to think about are frequency, power, and cost.

Higher frequency gives you a higher bit rate, but can be blocked by obstacles easily. More power makes your signal "louder" so it can travel farther, but is going to deplete your batter faster as a result. The rule of thumb is bitrate, range, cost—pick two.

1. The first thing to do is add decoupling capacitors to the Arduino ground and VIN, as this will be powered from battery. Also add capacitors to the power and ground for each of the servos, as shown in Figure 1-9.

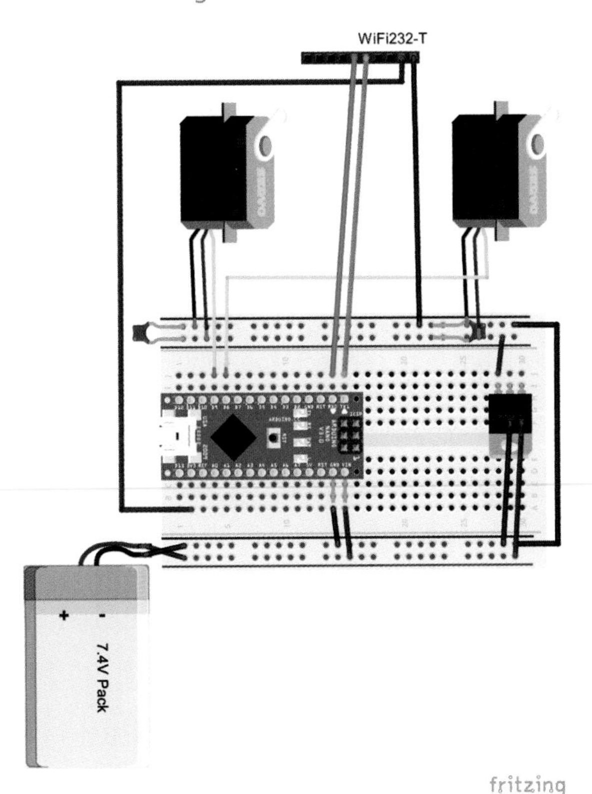

Figure 1-9 *WiFi wiring diagram*

2. The wireless module uses 2mm pitch headers, so it can't be plugged into the breadboard directly. Use the M-F jumper wires to connect the module to the breadboard instead. Add the WiFi module with the connections illustrated in Figure 1-10 and mapped in Table 1-3.

Table 1-3 *WiFi module connections*

WiFi232 Pin	Arduino Pin
1	GND
2	VCC (3.3v); *don't* plug this into 5V!
5 (RX)	Arduino (TX) Pin 1
6 (TX)	Arduino (RX) Pin 0

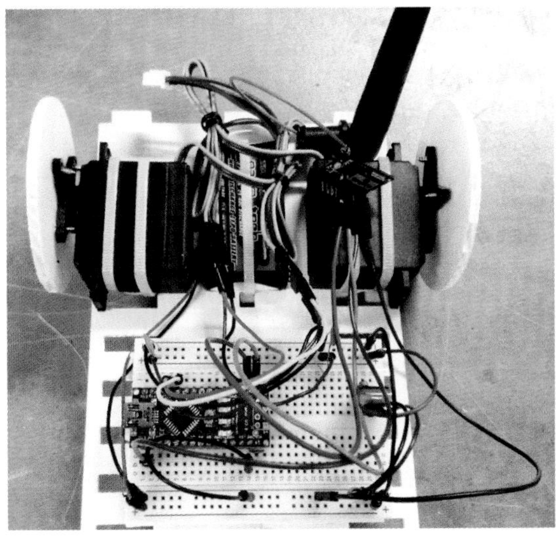

Figure 1-10 *Added wireless module*

3. Check, check, and triple check your wiring. There's no reverse or over-current protection on these modules, and applying power with too much voltage or GND and VCC back to front will quickly toast your WiFi module.

   These modules are designed to take an external antenna, so your range will be drastically reduced if you don't use it.

The antenna can be readily attached to the chassis with a nice lump of Blu-Tack or hot glue.

4. Finally, put a jumper wire from the battery power rail to the VIN pin on the Arduino. This will give the Arduino 7.4V, which is plenty for it to run.

That was easy; now for the software side.

Controlling the SimpleBot

Let's walk through the wireless portion in stages. When you remove wires, there's more complexity, so you need to make sure each element works before moving onto the next. If anything goes wrong, just backtrack a little and try again.

Test and configure the module

Once 3.3V is supplied, the WiFi module will power up. By default, it will be in Access Point mode. Wait about 10–15 seconds and then in your available WiFi networks on your computer you should see the "USR-WIFI232-T" SSID appear. Connect to this network and once connected, you can try two things to see if it's all working.

The module IP address is 10.10.100.254, and it has a DHCP server so it should assign you something in that range once you connect. Test pinging 10.10.100.254 and if you're getting responses like those shown in Example 1-3, move to the next step.

Example 1-3 *Checking connection to the module*

```
$ ping 10.10.100.254

PING 10.10.100.254 (10.10.100.254) 56(84) bytes of data.
64 bytes from 10.10.100.254: icmp_req=1 ttl=255 time=2.21 ms
64 bytes from 10.10.100.254: icmp_req=2 ttl=255 time=23.5 ms
64 bytes from 10.10.100.254: icmp_req=3 ttl=255 time=29.5 ms
^C
--- 10.10.100.254 ping statistics ---
3 packets transmitted, 3 received, 0% packet loss, time 5013ms
rtt min/avg/max/mdev = 2.215/20.281/29.502/9.622 ms
```

If this doesn't work, diagnose your network interface and make sure you're not using a static IP or some other configuration that overrides DHCP.

Now open a web browser and point it at 10.10.100.254. The username and password by default are both "admin." In this interface, you can configure the common aspects of the module itself. Try setting it to operate in STA+A mode, which means it operates as a WiFi station (connects to your WiFi network), but also operates as an access point if you need it (like when taking a SimpleBot to the park).

If you've ever configured a WiFi router, then everything will look pretty familiar. Supply the connection details for your network and get the module to connect. You should now see all of the connection information.

If you can't get to the wireless module using these methods for some reason, try connecting over serial directly. You'll need something like an FTDI or other serial ttl USB cable. Connect using a serial console at 115200 baud. The WIFI232-T user guide (http://bit.ly/19LX1ni) has more information on using AT commands.

Arduino setup

Because the WiFi module can work at a higher baud rate, the Arduino should talk at that speed, too. To change this, open the StandardFirmata sketch and then do a "Find" for the line that looks like this:

```
Firmata.begin(57600);
```

Change it to connect at 115200 instead:

```
Firmata.begin(115200);
```

That's all. Compile and upload the sketch to the Arduino.

stk_500 sync error?

Because you're using the Arduino hardware serial port, if you need to flash your Arduino, you'll need to remove the RX and TX wires connected to the WiFi module so you can talk to the Arduino over USB. If you get an stk_500 sync error, it's probably because you forgot to unplug the serial wires.

Network test

The next step is to make sure communication is occurring properly before trying to send messages from Johnny-Five and the application.

The WIFI232 module exposes TCP port 8899, and whatever is sent to and from the serial connection is sent through that TCP port. Messages will get passed like this:

```
PC <--> WiFi Network Interface (TCP 8899)
   <--> WIFI232 Network Interface (TCP 8899)
   <--> WIFI232 Serial Interface
   <--> Arduino Serial Interface
```

If you connect to the module using Telnet, you should be able to see any serial messages coming from the Arduino. Although most Firmata data is encoded, one message is in clear text—when it sends the name of the Firmata sketch being used to the receiver as part of startup.

On my networks, the module is using IP: 10.0.1.12 so telnet there on port 8899 with `telnet 10.0.1.12 8899`.

Windows doesn't usually install the Telnet client by default. You'll need to make your way to the Add/Remove Windows Components section of the Control Panel and install it.

This gives the following response:

```
Trying 10.0.1.12...
Connected to 10.0.1.12.
Escape character is '^]'.
```

That's good, because everything is connected. Hit the reset button on the Arduino so the sketch restarts. You should see the telltale Pin 13 LED blink sequence and within a few seconds something like this:

```
Trying 10.0.1.12...
Connected to 10.0.1.12.
Escape character is '^]'.
��yStandardFirmata.ino
```

Success! You just received a message from Firmata onto your PC without any wires. Quit Telnet (press Ctrl+] then type "quit"). You're now ready for the step you've been waiting for.

Controlling the SimpleBot wirelessly

Now that you have Firmata messages traveling over the network, all you have to do is get Johnny-Five to read and write them. The problem is that Johnny-Five assumes you have a Serial device connected—something like "/dev/ttyUSB0" or "/dev/tty.USBSerial." Instead, you have a network socket. One option is to go and write an IO interface for Johnny-Five, but that's overkill for a simple socket, and it's not like you're writing a new messaging protocol—you're just sending messages via the magic of WiFi rather than a pair of wires. Instead, follow the steps described here (for Mac/Linux, use socat, and for Windows, use VSPE):

Socat for Mac/Linux computers

Socat creates relays between two otherwise independent data channels. Knowing this, you can use pseudoterminals to allow you to create a "fake" serial terminal. A "faked" serial terminal can then be used by Johnny-Five and our application. You should be able to install socat on Linux from your distribution's package manager. On Mac OS X, you could use Homebrew (*http://brew.sh*).

Make a relay between a pseudoterminal that looks like a serial port and connect that to the TCP socket used to talk to the WIFI232 module. On my networks, the WiFi module is at 10.0.1.12—yours will be different, so just replace that. Here's how it will work:

```
PC <--> Pseudo terminal (~/dev/ttyV0)
   <--> TCP Socket (10.0.1.12:8899)
```

To create this route, use this command (remember to replace 10.0.1.12 with the correct address for your WiFi module):

```
socat -d  pty,nonblock,link=$HOME/dev/ttyV0
tcp:10.0.1.12:8899
```

The `-d` switch is a debugging parameter, so remove it if you don't want details, and use it up to four times for lots of messages if something doesn't work.

This command tells socat to create a pseudoterminal (`pty`). Don't block it (`nonblock`; you can use it from other processes) and `link` it to *$HOME/dev/ttyV0* (in this case that puts it at */home/ajfisher/dev/ttyV0*, but put it wherever you fancy). Socat then connects that terminal to a TCP connection at 10.0.1.12 using port 8899. Once socat is running, create another terminal window and use `screen` to connect to a fake serial connection:

```
screen ~/dev/ttyV0
```

Do the Arduino reset trick again and you will see the Firmata sketch name. If that all works, then you're ready to wirelessly control your SimpleBot. All you need to do is pass in the fake serial connection to the program you wrote before, and away you go. At this point, it will work just like it did before, but your SimpleBot will now be wireless:

```
node simplebot.js /home/user/dev/ttyV0
```

VSPE for Windows computers

On Windows, use a piece of software called Virtual Serial Ports Emulator (VSPE). This allows you to define a COM port and then create mappings for other endpoints such as to network services.

In the end, the messages should route like this:

```
PC <--> Virtual COM Port (COM10)
   <--> Bridge
   <--> TCP Client (10.0.1.12:8899)
```

Download and install the software and once you have it open create two devices:

Create the first one as a "Connector" from New Devices and set it to be COM10—this creates a COM port for Windows to talk to and will be what is passed to the Node.js application:

- TCP Client
- Host: 10.0.1.12 (or whatever your WIFI232 IP is)
- Port: 8899
    - The next one created is a "Bridge" type, which bridges between two data streams:
- Serial port
- Port: COM10 (type it in—it won't be in the drop-down as it's not "real")
- Speed: 115200

Save the config and press the "play" button to have it all run.

You should now just be able to run the SimpleBot application like before:

```
node simplebot.js COM10
```

Troubleshooting

The most common problems here are related to power or range causing resets.

If you don't have enough power or haven't used decoupling capacitors, your Arduino or WiFi module may reset when you do big direction changes on the motors. Check the circuit, make sure you put the capacitors in, and check the power levels on your battery pack.

If you are getting persistent resetting issues, you may be drawing too much current from your servos. If they are particularly heavy duty (high torque), then it's possible you're dropping the voltage too much for the Arduino. You may need a higher capacity battery designed for RC car use.

If you are going out of range of your WiFi network, you may get garbled messages and your SimpleBot stops responding. This is where you start thinking about bigger antennas, and other radio types in order to make the range even longer.

What's Next?

Over the course of this chapter, you've taken some cardboard and turned it into a robot. With just a basic robot made from some scrap and some inexpensive components, you've explored a range of topics:

- Differential drive
- Programming control systems
- Remote control

Now that you have your own SimpleBot working, you can take it in many interesting directions to learn even more robotics concepts. Here are some things you might want to try:

- Add autonomy and collision avoidance.
- Add a Raspberry Pi running Node.js to do all of the processing and logic on the robot and make it even more autonomous.
- Use a Raspberry Pi tethered to your mobile phone and make a remote web-controlled bot that can drive anywhere.
- Use some reflection sensors to get your SimpleBot to follow lines.
- Add some encoder wheels so you can measure distance of travel/speed, and turn more accurately.
- Add a light sensor and program your SimpleBot to be attracted to or run away from light.
- Program your SimpleBot to spin around and light up some LEDs whenever you get a tweet or someone sends you an email.
- Explore different drive mechanisms using more wheels or even omniwheels for a SimpleBot that can change direction without spinning on the spot!

2 TypeBot

By Bryan Hughes

TypeBot is a robot that can type on a keyboard for you. It was first created during a hack day at JSConf 2013 in Florida after a team of three was inspired by a talk given by Raquel Velez on robotics. Some members of the team had prior experience with robotics dating back to the

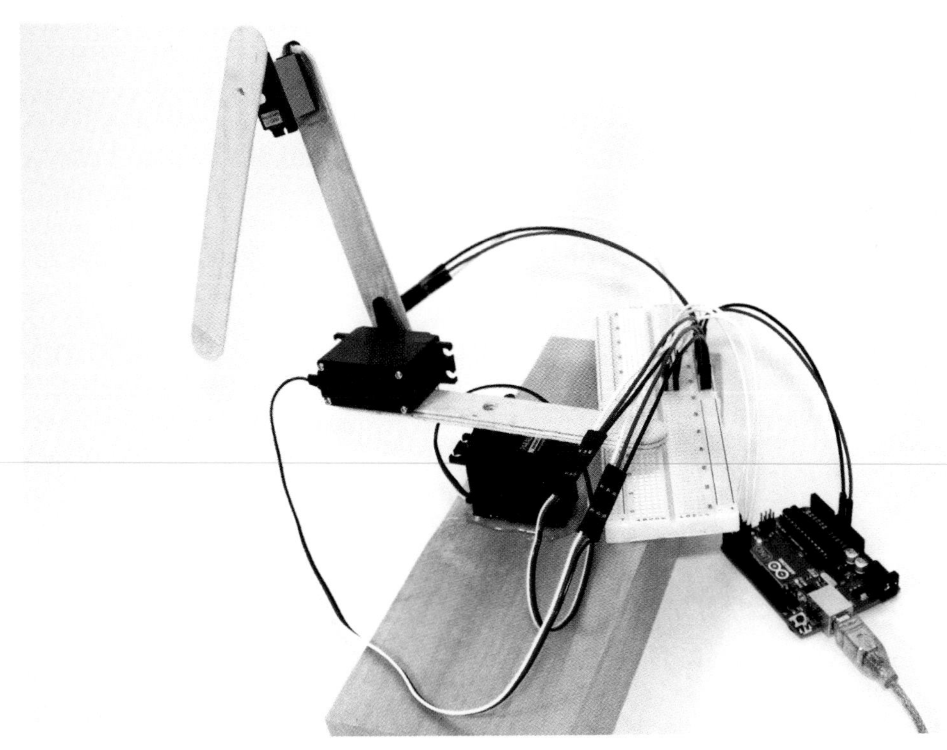

Figure 2-1 *Completed arm*

early 2000s, but none had ever used JavaScript for robotics before. The ease with which JavaScript can be used to create robots was a major "aha!" moment for the team, as we hope it will be for you.

The Arduino Uno is an ideal microcontroller for TypeBot (see Figure 2-1) because it has plenty of pins for controlling servos. Johnny-Five is also quite mature on the Uno these days, and has the best support of all its supported platforms.

So why create a typing robot? Because we can! Also, it looks really cool.

Bill of Materials

To build TypeBot, you will need the materials listed in Table 2-1.

Table 2-1 *Bill of materials*

Count	Part	Estimated Price	Part Numbers
1	Arduino Uno	$24.95	MS MKSP99; AF 50; SF DEV-11021
3	High-torque servo	$12.95	SF ROB-11965, AF 155
1	Breadboard	$4.95	SF PRT-12002, AF 64, MS MKKN2
14	Jumper wires (male to male)	$4.95	SF PRT-11026, AF 758, MS MKSEEED3
6	Popsicle sticks	$4.50	AZ B0033F7YQW
1	2 × 4 wood base	$12.50	AZ B00IZ94BG2

If you have an Arduino servo shield, you can use that instead of the breadboard and jumper wires. The Arduino servo shield is an expansion board that snaps onto your Arduino. This board provides headers that you can plug your servos into directly, without the need for manually wiring everything up. You may still need a few jumper wires if the servo cables are too short, however.

You will also need the following tools:

- Hot glue gun
- Small Phillips screwdriver
- Drill

The drill is used to cut holes in Popsicle sticks, so it doesn't need to be powerful. You can use a knife or screwdriver in a pinch if you don't have a drill, but they are much more difficult to work with. If you use a knife, you can make a hole by poking it into the stick and rotating it with your fingers. Rotating the knife will slowly "drill" out the stick. Using a knife in this manner will add wear and tear on the blade, so don't use an expensive knife.

Understanding Your Servomotors

We will be using servomotors (servos for short) as the "joints" in the arm. There are multiple types of servos, and it's important to understand the differences between them. Servos can generally be divided into two categories: *standard* and *continuous*.

A standard servo is a servo that has a limited range of motion, measured in degrees. With a standard servo, you send a signal to the servo that is translated to a *position*. These types of servos are commonly used in remote-controlled (RC) cars to control the steering. Standard servos can typically set any position within a 180° arc, although 90° servos are also common.

A continuous servo can spin freely. With a continuous servo, you send a signal to the servo that is translated to a *speed*. These types of servos are commonly used in RC cars to control the drive wheels.

So what about this signal that is used to control a servo? All servos are driven by a *pulse-width modulated* signal, or *PWM*. Servos expected pulses to come every 50 milliseconds. The width of the pulse determines the position on a standard servo, and the speed on a continuous servo. A width of 5 ms is all the way to the left, and a width of 15 ms is all the way to the right.

Sometimes you will come across a servo that doesn't quite match these numbers, though, so you may need to tweak the numbers a little bit.

This project requires three 180° standard servos, although two of the three servos can be 90° servos if that's all you have. The rest of this chapter will assume that you have 180° servos.

Anatomy of a Robot Arm

We all know how an arm works, right? There's not much to think about when they are being controlled by a human brain. Our brains are wonderfully complex constructs that take all the fidgety little details out of controlling arms. Picking up an object is very easy for us. We just kinda think about the gist of what we want to do and our nervous system does the rest for us.

Robotic arms don't have anything nearly as powerful as a human brain controlling them, however. This means that we have to handle all the nitty-gritty details ourselves. The layout of the arm we choose will determine what type of constraints there are on movement, as we shall see.

Arm Layout

TypeBot will be using an arm with three servos to act as the joints, which we will call the "shoulder," "elbow," and "wrist."

The elbow and wrist will be connected via Popsicle sticks. The servos are installed 180° apart from each other so that they can move opposite of each other. Moving the servos at the same rate will cause the tip of the finger to move in a straight line. Figures 2-2 through 2-4 show the various stages of movement.

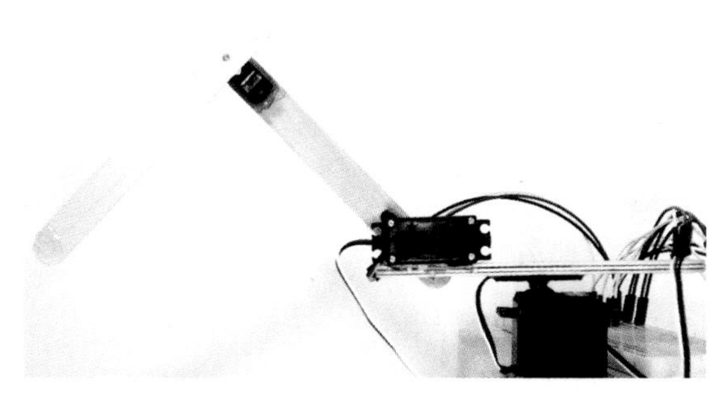

Figure 2-2 *Moving in a line, extended*

Figure 2-3 *Moving in a line, contracting*

Figure 2-4 *Moving in a line, contracted*

This structure is then attached to the shoulder servo and with the arm segment rotated 90° from the rest of the arm. This third servo is used to set the angle of the line along which the other two servos expand and contract. This allows the tip to move to any location on the keyboard.

Arm Constraints

Most robots are pretty clumsy, and we don't want one to knock into our keyboards (especially if those keyboards are built into our nice laptops). What we need is to orchestrate the movement of the arm, just like you would orchestrate a complex animation in a web page. If you think about it, this is the ultimate animation, because it's animating something in real life.

We could make use of some awesome math, called *constraint programming*, to create an optimal animation that is guaranteed to never hit anything. We could also just come up with something quick and dirty and use the extra time to show off TypeBot to all our friends, which sounds a lot more fun!

So what sorts of constraints do we have to worry about? Let's assume that you will be using TypeBot on a laptop keyboard, because it has a few extra constraints compared to a standalone keyboard.

The first constraint is that we don't want to hit the laptop screen by overextending the arm. The maximum amount we can extend the elbow joint varies based on the angle of the base servo. This means we can't just say "don't extend past *x* degrees." Fortunately, this constraint is directly related to how far we need to extend the arm to reach the keys themselves. If you look down at your keyboard on your laptop and pretend the center of your trackpad is the center servo, you can see how much closer the 6 key is than the 1 key to that point. This means that the arm needs to extend and contract as it moves left and right along the keyboard. If we don't time the extension of the arm and the rotation of the shoulder, it will hit the screen.

The second constraint is that we don't want to drag the arm across the keyboard as we move between keys. This isn't a Swype virtual keyboard after all! We have to ensure a few things here. When the arm is rotating, we need to make sure the finger hovers safely above the keyboard. We also need to make sure that, when we press a key, the arm lowers onto the key cleanly (i.e., perpendicular to the keyboard). If we don't lower the finger cleanly, we risk sliding off of the key.

How do we solve those constraints? It wouldn't be any fun to tell you now, so read on!

Building the Hardware

Now it's time to get our hands dirty and build the robot arm!

The Base and Shoulder

Start by getting the wood base and one of the servos. We are going to connect the servo to the base using hot glue. Don't worry, hot glue peels off easily if you want to reuse the servo later. You'll need to follow these steps:

1. Put a layer of glue on the bottom of the servo. Make sure to use plenty of glue because we don't want any air gaps between the servo and the base.

2. Press the servo against the base so that the servo axis is about 1/3 of the way from one of the edges.
3. Quickly position the servo how you would like it.
4. Let it sit for about 30 seconds to solidify.

Next, we need to reinforce the servo bond. Remember how we are using the heavy wood base to stabilize the arm as it moves? All of that force will now be on this glue joint, so we need it to be strong. Add a ring of hot glue between the servo and the base, as if you are caulking a shower or trimming the icing around the base of a cake. The attached servo should look like Figure 2-5.

Figure 2-5 *Base with shoulder*

The Elbow

Next up, we are going to build the first arm segment and the elbow joint. This segment needs to be strong because it will be supporting the entire weight of the arm. This segment is also, unfortunately, angled such that the force of the arm is along the flat, flexible side of the stick. Here are the steps you should follow:

1. To make sure the arm is strong, get out three Popsicle sticks and hot glue them together. Once again, make sure to get plenty of hot glue between the sticks so that there is no air gap.
2. Allow the glue to dry for at least 30 seconds.
3. Drill a hole through the center of the stick set. This hole needs to be big enough to allow the screw for the servo arm to fit through it. A 1/4" drill bit should be big enough to allow the head of the screw to fit through it.
4. Take out the servo arm and add a light layer of hot glue on it. Try not to get glue over the center hole where the screw fits through if you can help it.
5. Attach the servo arm to the stick set such that the hole in the arm lines up with the hole you just drilled:
    a. If you got glue into the hole in the middle of the stick, you can use the servo arm screw to drill it by screwing it into the glue.
    b. Keep screwing it until the glue becomes loose, and push the screw out the other end using your screwdriver.
6. Take out another servo to serve as the elbow joint.
7. Add enough hot glue to prevent air gaps to one end of the stick group from the previous step. The glue should be placed on the *opposite* side of the stick to which the servo arm is glued.
8. Press the servo to the stick set *on its side*, such that the axis is perpendicular to the end of the stick, as shown in Figure 2-6.

Figure 2-6 *Elbow joint and arm segment*

Now it is time to connect the elbow joint to the base. Connect the white wire on the shoulder servo plug to pin 3 of the Arduino, the red wire to 5V, and the black wire to GND. To make sure that you align the arm properly; you are going to write a tiny Johnny-Five program to keep the shoulder servo centered.

Make sure your Arduino has been prepared as directed in "Arduino" and is connected to the computer or device on which you'll be running Johnny-Five. If you have any trouble running this project, see "Installing Johnny-Five".

Let's start by creating a folder called *TypeBot/*. Open a Terminal and navigate to this folder. For example, if you are on OS X or Ubuntu and created this folder in your *Documents/* folder, type:

```
cd ~/Documents/TypeBot
```

If you're on a Windows, OS X, or Linux system that doesn't already have Node installed, see "Installing Node.js".

Once inside of this folder, install Johnny-Five by typing:

```
npm install johnny-five
```

Create a new text file inside of this folder called *align.js*, and paste the following code into it:

```
var five = require("johnny-five");
var board = new five.Board();

board.on("ready", function() {
  new five.Servo(3).center();
});
```

Run the program with the following command:

```
node align.js
```

All source code for the examples in this book can be found on GitHub (*https://github.com/rwaldron/javascript-robotics*).

You should leave the program running while you connect the arm to make sure the servo stays centered.

Connect the elbow joint to the shoulder servo by sliding the servo arm on the bottom of the stick set onto the exposed servo gears. The stick should be perpendicular to the base, with the elbow joint hanging over the edge of the base. Once in place, screw the arm into the shoulder servo to secure it. The result should look like Figure 2-7.

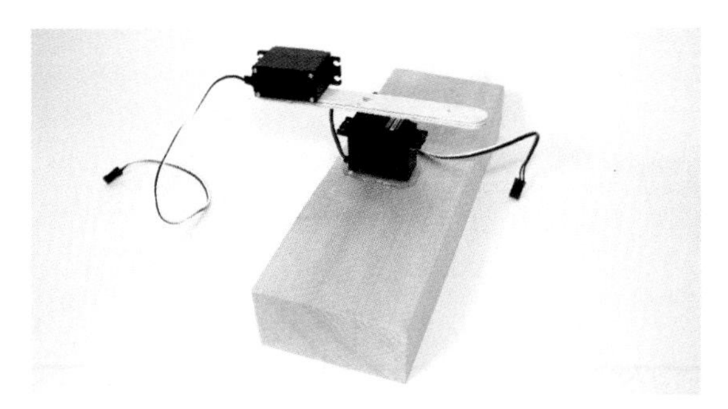

Figure 2-7 *Attachment of the elbow joint servo*

The Wrist

The next arm segment is similar to the first arm segment, but with a few changes. This segment will only be built with two Popsicle sticks, instead of three. We can get away with fewer sticks here because this segment doesn't need to support as much weight. You might even be able to get away with a single stick. The second stick can be helpful in reducing flex, but it does add weight, so it's up to your personal preference. You'll need to follow these steps:

1. Start by gluing two sticks together, as before.
2. Drill a hole about one inch from one end of the stick. The servo arm should be as close to the edge as possible, without hanging over.
3. Glue the servo arm to the stick using the same techniques as before.
4. Get your final servo out.
5. Add glue to the opposite side and end of the stick that the servo arm is glued to.
6. Press the servo to the stick such that the servo axis is pointing up from the stick, as shown in Figure 2-8.

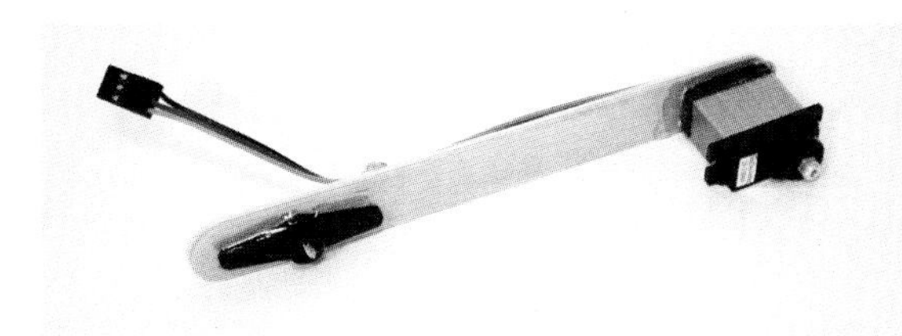

Figure 2-8 *Attachment of the wrist servo*

Now it's time to connect the wrist arm to the elbow servo. We will use the exact same technique used to connect the elbow arm to the base, except that you should connect the elbow servo to the Arduino, instead of the shoulder servo. Connect the elbow arm so that it is pointing straight up into the air, as shown in Figure 2-9.

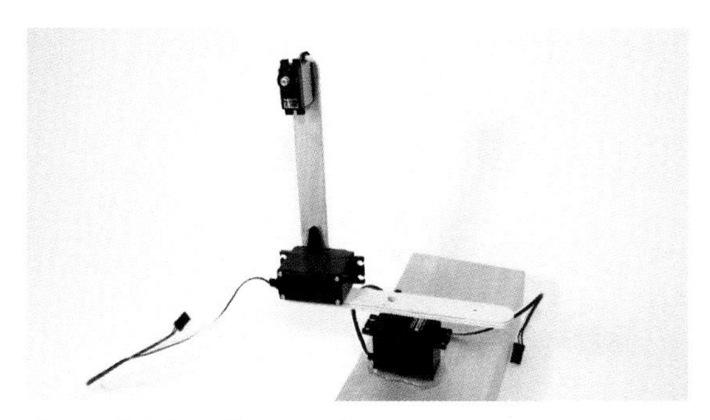

Figure 2-9 *Installation of the wrist segment*

The Finger

Now it's time to build the last segment of the arm, which also happens to be the easiest! This arm segment doesn't need much strength, so we will only use a single Popsicle stick. Start by drilling a hole about one inch from the end, like we did for the wrist arm. Glue the servo arm to the stick, as before.

The one new thing we will do here is coat the other end of the arm in hot glue. We coat the end of the stick for two reasons. The first is that we don't want the wooden end of the stick to scratch up our keyboard. The other reason is that hot glue will give the stick a lot more grip. This will prevent the arm from sliding off of a key when we are trying to press it. The finger should look like Figure 2-10.

Figure 2-10 *Finger arm segment*

Now it's time to connect the finger. Disconnect the elbow servo from the Arduino and connect the wrist servo. Run the program to center the servo, and attach the finger such that it is perpendicular to the wrist arm and parallel to the ground. With this step, you now have a fully constructed arm! There is only one thing left to do.

The Brains

The final build step is to connect all of the servos to the Arduino and secure the wiring. It is recommended that you glue the wrist servo wire to the end of the wrist arm, as shown in Figure 2-11. This will help keep the wrist servo wire from getting caught while the arm is moving.

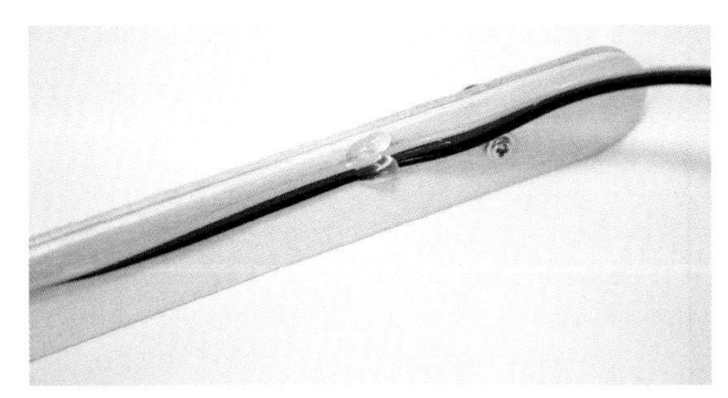

Figure 2-11 *Securing the wrist wire*

Connect the servos and the Arduino to the breadboard using *jumper wires* or *servo wire extensions*, as shown in the Fritzing diagram (*http://fritzing.org*) (Figure 2-12).

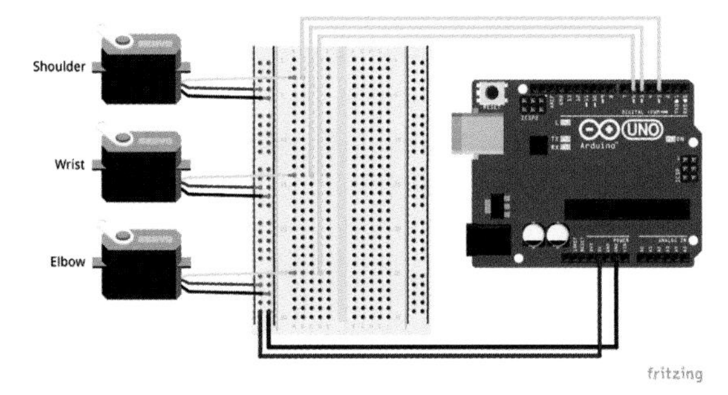

Figure 2-12 *Wiring diagram*

You should make sure that there is plenty of slack in the wires coming off of the arm so that there is enough room for the arm to move. The final arm should look like Figure 2-13.

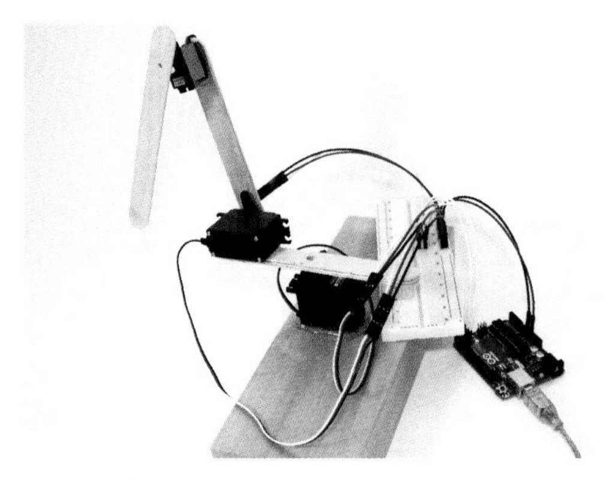

Figure 2-13 *Complete arm*

Now let's make it type!

Writing the Software

We are going to break the software into two parts: servo control and sequence management. The servo control module adds an abstraction layer on top of the Johnny-Five servo APIs. We use this type of abstraction to make it easy to coordinate the movements of multiple servos.

Creating the Project Files

First, you need to create the files that will contain the code:

1. Open a terminal and navigate back to the *TypeBot/* folder that you created earlier.
2. Create two files inside of this folder: *typebot.js* and *servocontrol.js*. These files will contain the source code for the two modules.

Controlling the Servos

First, create the servo control module. Start by adding the boilerplate shown in Example 2-1.

Example 2-1 *Servo control boilerplate*

```
var five = require('johnny-five');
var positions = {};
var servos = {};
```

```
var opts;

// Initializes the servo control module
function init(board, options, callback) {
}

// Move the servos
function move(destinations, callback) {
}

// Export the public methods
module.exports = {
  init: init,
  move: move
};
```

This creates a module with two methods, `init` and `move`, that *typebot.js* will call. It also imports the Johnny-Five library and creates some global state variables.

The initialize method is responsible for creating the Johnny-Five servo instances and initializing the global state information. The `init` parameters contain three pieces of information—the individual servo configurations, the servo rotation rate, and the settle time:

- Each servo configuration is identified with a name and will contain the pin number, the starting position, and whether or not to invert the servo angles (more on this momentarily).
- The servo rate will specify the *maximum* speed that a servo can rotate at, in degrees per millisecond, although it may spin slower.
- The settle time is used to add an additional delay after the servos have finished moving. When the arm moves, it shakes around. Once it stops, it typically continues to shake for a brief period of time. The settle time gives the arm a chance to stop shaking before moving on to the next step.

Here is what the configuration object looks like with everything included (you don't need to add this to the code listing now):

```
{
  servos: {
    shoulder: {
      pin: 3,
      startPosition: 90,
      isInverted: false
    },
    elbow: {
      pin: 6,
      startPosition: 60,
      isInverted: false
    },
    wrist: {
      pin: 5,
      startPosition: 30,
      isInverted: true
    }
  },
  rate: 0.01,
  settleTime: 250
}
```

The complete `init` method is shown in Example 2-2.

Example 2-2 *The completed init method*

```
// Initializes the servo control module ❶
function init(board, options, callback) {

  // Store the options for use with move()
  opts = options;

  // Initialize the servos ❷
  for (var servo in options.servos) {

    // Alias the servo config for easy access
```

```
    var servoConfig = options.servos[servo];

    // Store the start position as the current position
    positions[servo] = servoConfig.startPosition;

    // Create the servo instance
    servos[servo] = new five.Servo({
      pin: servoConfig.pin,
      isInverted: servoConfig.isInverted
    });

    // Move to the servo to the starting position
    servos[servo].to(positions[servo]);
  }

  // Wait for the servos to move to their starting positions ❸
  setTimeout(callback, 1000);
}
```

❶ Let's start by modifying the init function to store the options. This way we can use them later in the `move` function.

❷ Next, let's initialize the servos themselves. This code will loop through each servo in the `servos` object and initialize it by performing the following steps:

1. Store the staring position in the global `positions` variable.
2. Instantiate the Johnny-Five servo instance and store it in the `servos` global variable.
3. Move the servo to the starting position.

❸ Finally, let's add a timeout to give the servos time to move to their starting positions. This line of code will call the callback after 1 second, telling *typebot.js* that initialization is complete.

Now it's time to implement the `move` method. This method will take in a set of destinations for multiple servos and make sure they move in a coordinated manner. We will calculate the speed that each servo needs to rotate at to make sure they all arrive at their destinations at the same time. The complete `move` method is shown in Example 2-3.

Example 2-3 *The completed move method*

```
// Move the servos ❶
function move(destinations, callback) {

  // Find the largest servo angle change
  var largestChange = 0;
  for (var servo in destinations) {
    var delta = Math.abs(destinations[servo] - positions[servo]);
    if (delta > largestChange) {
      largestChange = delta;
    }
  }
```

```
  // If none of the servos need to move, short-circuit here ❷
  if (largestChange === 0) {

    // We still need to call the callback, but we want the callback to always be
    // asynchronous, so we use process.nextTick to call it asynchronously.
    // For more information on why this is a good thing, read:
    // http://nodejs.org/api/process.html#process_process_nexttick_callback
    process.nextTick(callback);
    return;
  }

  // Calculate how long we should take to move, based on the largest ❸
  // change in angle. This means that only this one servo will move at full
  // speed.  All of the other servos will move at a slower rate so that all
  // servos finish at the same time.
  var duration = largestChange / opts.rate;

  // Move the servos to their destinations ❹
  for (servo in destinations) {
    positions[servo] = destinations[servo];
    servos[servo].to(destinations[servo], duration);
  }

  // Wait until we are done and call the callback ❺
  setTimeout(callback, duration + opts.settleTime);
}
```

❶ We start by finding out which servo needs to move the furthest. We loop through each change and find out which one needs to move the furthest.

❷ To make the code more efficient, let's check to see if none of the servos need to move. Notice that we are not calling the callback directly, but are wrapping it in a `process.nextTick` call. If we didn't do this, the callback would be called synchronously if there are no changes, but asynchronously if there are changes. This can introduce subtle bugs that are hard to find. For more information on this issue, check out the link in the code comments.

❸ We calculate how long it will take the servo with the greatest delta to move to its new destination.

This code calculates how long it will take this servo to move to its position while moving at the rate supplied in the options. We use the duration we just calculated to move *all* of the servos, which means the other servos will move at a slower rate. Remember the laptop screen constraint from the beginning of this chapter? This is how we handle it! It turns out that the arm can only hit the screen when the arm starts at keys near the middle top of the keyboard (i.e., t, y, etc.) and moves to keys near the upper edges of the keyboard (i.e., q, p, etc.). Moving the other servos at slower rates prevents the elbow and wrist from rotating too fast and hitting the screen, because the shoulder will always be rotating faster than the elbow and wrist in these cases.

❹ With that out of the way, let's move the arm. We loop through all of the servos, update their positions, and move them.

❺ At this point in time in our code, we have done everything we need to do to tell the servos to move to their positions. Now we just need to wait for the hardware to do its thing. We simply wait for `duration` milliseconds to pass, and then wait for `settleTime` milliseconds to pass, and call the callback.

Initialization

Now that we have our servo module, let's initialize it in the main module. Open *typebot.js* and add the following code:

```
var five = require("johnny-five");
var servocontrol = require("./servocontrol");
```

Next, let's create a sequence of keys to type. The sequence we want to type on our first try is quite obvious, of course:

```
// This is the sequence of keys that are
// pressed. Each element in the array needs
// to correspond with an entry in the
// KEYS object.
var SEQUENCE = ["h", "e", "l", "l", "o", "w",
"o", "r", "l", "d"];
```

Now, let's create the servo options. First, we define the servo rotation rate to be 0.05 degrees per millisecond, which is a little over 8 RPM. We want to start with a low servo rotation rate as we finetune the system. Once everything is working at this speed, we'll ramp it up to 0.2 degrees per millisecond, or 33 1/3 RPM (my favorite rotational speed!):

```
// This is the rate that the servos should
// rotate at, measured in degrees per
// millisecond. 0.05 is a good value to
// start with.
var SERVO_RATE = 0.05;
```

Next, we define the settle rate to be one quarter of a second. If you find that the arm gets more unstable as it types, try increasing this value:

```
// This is the delay between movement steps.
// A delay gives everything a chance to
// settle and helps prevent overdriving the
// arm.
var STEP_SETTLE_TIME = 250;
```

Now let's create the servo configuration:

```
// This is the basic servo config.
// Each key-value pair names and defines a ser
vo.
// The port is the pin number on the Arduino
header,
// and the defaultPosition is the position that
// the servo will start at when the app boots
up.
var SERVO_CONFIG = {
  servos: {
    shoulder: {
      pin: 3,
      startPosition: 90,
      isInverted: false
    },
    elbow: {
      pin: 6,
      startPosition: 60,
      isInverted: false
    },
    wrist: {
      pin: 5,
      startPosition: 30,
      isInverted: true
    }
  },
  rate: SERVO_RATE,
  settleTime: STEP_SETTLE_TIME
};
```

We configure the servos to start out in the position shown in Figure 2-14.

Figure 2-14 *Starting position*

If you run the code and it doesn't match this position, make sure you have the `isInverted` flags set appropriately. It's OK if your angles don't match the figure exactly, just as long as they are close.

The last piece of configuration we need to create is a mapping between keys and servo angles:

```
// These are the keys on the keyboard,
// and the position of each servo when
```

```
// they are *pressing* the key
var KEYS = {
  a: { shoulder: 125, elbow: 19, wrist: 87 },
  b: { shoulder: 88, elbow: 21, wrist: 62 },
  c: { shoulder: 105, elbow: 21, wrist: 65 },
  d: { shoulder: 114, elbow: 21, wrist: 77 },
  e: { shoulder: 114, elbow: 19, wrist: 87 },
  f: { shoulder: 107, elbow: 21, wrist: 74 },
  g: { shoulder: 100, elbow: 21, wrist: 72 },
  h: { shoulder: 92, elbow: 21, wrist: 70 },
  i: { shoulder: 81, elbow: 20, wrist: 79 },
  j: { shoulder: 84, elbow: 21, wrist: 70 },
  k: { shoulder: 77, elbow: 21, wrist: 71 },
  l: { shoulder: 69, elbow: 21, wrist: 73 },
  m: { shoulder: 70, elbow: 21, wrist: 65 },
  n: { shoulder: 78, elbow: 21, wrist: 63 },
  o: { shoulder: 75, elbow: 20, wrist: 81 },
  p: { shoulder: 68, elbow: 20, wrist: 83 },
  q: { shoulder: 124, elbow: 16, wrist: 98 },
  r: { shoulder: 108, elbow: 20, wrist: 83 },
  s: { shoulder: 120, elbow: 20, wrist: 82 },
  t: { shoulder: 102, elbow: 20, wrist: 81 },
  u: { shoulder: 88, elbow: 21, wrist: 78 },
  v: { shoulder: 97, elbow: 21, wrist: 63 },
  w: { shoulder: 119, elbow: 18, wrist: 92 },
  x: { shoulder: 113, elbow: 21, wrist: 68 },
  y: { shoulder: 95, elbow: 20, wrist: 79 },
  z: { shoulder: 120, elbow: 21, wrist: 72 }
};
```

Where did these numbers come from? A small program (*http://bit.ly/1wDK7CZ*) was written that performs some trigonometry to estimate the position of keys. Note that this program can only generate estimates because of subtle variations in servo width/height, slightly misaligned screw holes, servos that aren't exactly 180°, and so on.

The last step is to initialize Johnny-Five and the servos:

```
function run() {
}

var board = new five.Board();
board.on("ready", function() {
  servocontrol.init(board,
                    SERVO_CONFIG, run);
});
```

Now it's time to run the code. Make sure that all of the servos are properly connected to your Arduino, and that the Arduino is plugged into your laptop. Run `node typebot.js` and the arm should move into its starting position!

Sequencing a Key Press

Now that we have our servo module, how do we actually use it to press a key given the constraints from the beginning of the chapter? We are going to use a state machine to break a key press into three steps:

1. Positioning the finger directly above the key.
2. Moving the finger down until the key is depressed.
3. Raising the finger back to the position the finger was in at the end of step 1.

These three steps can be represented by the state machine shown in Figure 2-15.

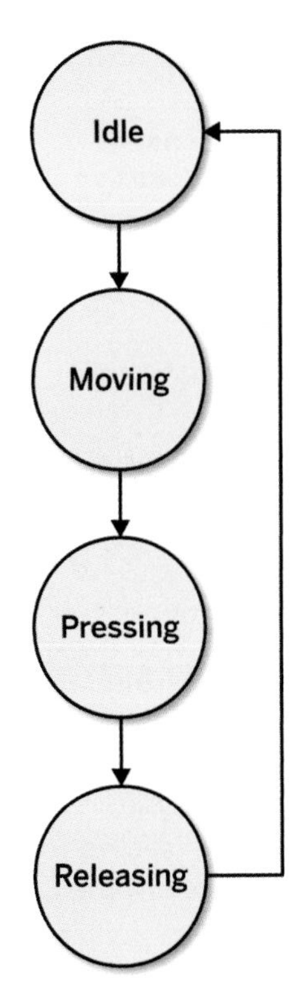

Figure 2-15 *State Diagram*

We can represent this state machine with an event loop and a switch statement. You should now add the code shown in Example 2-4 to the run method.

Example 2-4 *The run method*

```
function run() {

  // Define the states ❶
  var STATE_IDLE = 0;
  var STATE_MOVING = 1;
  var STATE_PRESSING = 2;
  var STATE_RELEASING = 3;

  // State machine information ❷
  var sequencePosition = -1;
  var state = STATE_IDLE;
  var key;

  function tick() { ❸
    switch(state) {

      case STATE_IDLE:
        break;

      case STATE_MOVING:
        break;

      case STATE_PRESSING:
        break;

      case STATE_RELEASING:
        break;
    }
  }

  // Kickstart the event loop
  tick();
}
```

❶ Here we define four states: idle, moving, pressing, and releasing.

❷ To keep track of the state machine, we use three global variables: `sequencePosition`, `state`, and `key`. `sequencePosition` keeps track of which key in the sequence we are currently pressing. `state` keeps track of which state within the sequence we are currently in. Finally, `key` is an alias for the current key entry in the `KEYS` array.

❸ We also create a function called `tick` that processes a single trip through the event loop. This method will be called four times for each key, once per state. Each state is responsible for transitioning to the next event. The code you see here is just the stub for these `case` statements. You'll see the full definition of each statement next.

The application starts out in the idle state. When the loop ticks, the idle state transitions to the moving state with the following code, which replaces the previous definition of case STATE_IDLE:

```
case STATE_IDLE:

  // Get the next key ❶
  sequencePosition++;
  key = KEYS[SEQUENCE[sequencePosition]];
  if (!key) {
    process.exit();
  }
  console.log('Typing key ' +
    SEQUENCE[sequencePosition]);

  // Move the arm to resting
  // above the key ❷
  state = STATE_MOVING;
  servocontrol.move({
    shoulder: key.shoulder,
    elbow: key.elbow + 10,
    wrist: key.wrist - 5
  }, tick);
  break;
```

❶ The first step is to get the next key from the sequence and store it. If there is no key, then that means we are finished typing and can exit.

❷ Then, we change the state to moving and tell the servo module to move the servos. We add an offset to the elbow and wrist so that they hover above the keyboard by about an inch. This prevents the arm from dragging along the keyboard. Note that we supply the `tick` method as the callback to the servo control's move method. This way, we will tick through the next loop in the event loop once the arm is finished moving.

The next state, moving, looks a lot like the second half of the idle state. The only difference is that we remove the offset so that the arm moves downward to press the key. These offsets were chosen because they cause the fingertip to move almost directly downward, which is one of the constraints from earlier. The following replaces the previous definition of case STATE_MOVING::

```
// Press the key
case STATE_MOVING:
  state = STATE_PRESSING;
  servocontrol.move({
    elbow: key.elbow,
    wrist: key.wrist
  }, tick);
  break;
```

The release state simply reverses the action taken in the previous state:

```
// Release the key
case STATE_PRESSING:
  state = STATE_RELEASING;
  servocontrol.move({
    elbow: key.elbow + 10,
    wrist: key.wrist - 5
  }, tick);
  break;
```

The final state is really simple. It changes state back to the idle state so that we can loop through again:

```
// Change to the idle state and pump the event
loop
case STATE_RELEASING:
  state = STATE_IDLE;
  tick();
  break;
```

In this state, we have to call `tick` directly because we aren't telling the servo to move. The complete `tick` method should now look like Example 2-5.

Example 2-5 *The final tick method*

```
function tick() {
  switch(state) {

    case STATE_IDLE:

      // Get the next key
```

```
        sequencePosition++;
        key = KEYS[SEQUENCE[sequencePosition]];
        if (!key) {
          process.exit();
        }
        console.log("Typing key " + SEQUENCE[sequencePosition]);

        // Move the arm to resting above the key
        state = STATE_MOVING;
        servocontrol.move({
          shoulder: key.shoulder,
          elbow: key.elbow + 10,
          wrist: key.wrist - 5
        }, tick);
        break;

      // Press the key
      case STATE_MOVING:
        state = STATE_PRESSING;
        servocontrol.move({
          elbow: key.elbow,
          wrist: key.wrist
        }, tick);
        break;

      // Release the key
      case STATE_PRESSING:
        state = STATE_RELEASING;
        servocontrol.move({
          elbow: key.elbow + 10,
          wrist: key.wrist - 5
        }, tick);
        break;

      // Change to the idle state and pump the event loop
      case STATE_RELEASING:
        state = STATE_IDLE;
        tick();
        break;
    }
  }
```

Running for the First Time

Are you ready to watch the arm type? Let's do it! To start with, you should run the arm *without* a keyboard below it just in case the arm goes haywire for some reason. Connect all of the hardware to your computer, and run `node typebot.js`. The arm should go about air-typing a message.

If nothing seems terribly wrong, it's time to try typing on a real keyboard. Place the keyboard under the arm, positioned such that the center of the shoulder servo is lined up with the B key. Run `node typebot.js` again and watch it type.

If you are *extremely* lucky, it will work the first time. More than likely, however, it probably won't get a single key right. This is OK, because this is where we fine-tune the arm.

Fine-Tuning the Arm

The easiest way to fine-tune a key is to comment out the releasing state case entry. This way, the arm will initialize, move above the first key, press the key, release the key, and stop. You can then set the letter you are interested in testing as the first entry in the sequence. Run through each key and adjust the angles until the key is being pressed properly. You may need to adjust the arm in all three directions to ensure that there is a good key press.

Once you are happy with how all of the keys are being pressed, reenable the releasing state case entry. Now TypeBot should be able to type out "helloworld".

Now it's time to crank it up to 11! Start increasing the speed in increments of 0.05. Run TypeBot with each increment and watch how it performs. As you increase the speed, small errors in the positions tend to become more pronounced. Fine-tune these keys using the method described earlier until it works at the new speed. Rinse and repeat until you have increased the speed to 0.2. Now you should be flying!

If you are feeling adventurous, you can keep trying to increase the speed, but be warned that you will start to run into physical limitations of the arm itself. It is entirely possible to drive the arm so fast that it tears itself apart!

What's Next?

Now that you have a functioning TypeBot, what else can you do with it?

TypeBot was originally created to type on a keyboard, but what about playing a piano? If you strengthen the arm, it should be able press a piano key without difficulty! If you string a few arms together, you can play a (slow) piano piece.

The current implementation requires a lot of fine-tuning to get it to work properly. A smarter robot would calibrate itself. You can add a camera to the end of the finger that runs optical character recognition (OCR) on the video feed to find letters on its own. The "node-tesseract" NPM module provides OCR capabilities in Node.js.

Too easy for you? Here's something pretty complex. It's a bit redundant having to type in the message that TypeBot will type. Wouldn't it be more fun to tell TypeBot what to type simply by talking to it? You can do just that with the Web Speech API (*http://bit.ly/1AYW0yl*). You will need to add a web server to TypeBot that serves up a web page to handle the speech-to-text aspects and then sends the text back to TypeBot telling it what to type.

Getting Started with NodeBoats

3

By Sara Gorecki

At JSConf 2014, NodeBoats was kicked-off with a full-day workshop: robots + water = fun! By the end of the day, participants had their boats sailing across the hotel pool, controlled from their laptops. I was part of a team of four that helped guide attendees through the process of creating their own seaworthy robots. This chapter will show you how to make one, too, as shown in Figure 3-1.

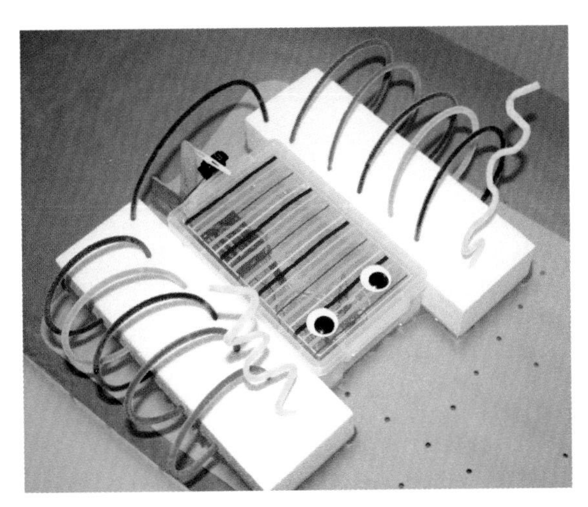

Figure 3-1 *NodeBoat*

One of the challenges of creating a boat is physically untethering your hardware from your computer. After all, you don't want to have to bring your laptop into the pool with you, or have your boat limited by the length of your USB cable. To this end, we'll use the *Spark Core* to control our boat. See "Spark WiFi Development Kit" for more on Spark Core.

As you can see in Figure 3-1, the sealable plastic container will be the hull of your boat, so make sure it's watertight! You want your breadboard and battery holder to fit comfortably inside, side by side. The boat pictured was made out of a 8.5″ × 5.5″ × 2.5″ plastic pencil case with a cover that snaps closed (we purchased ours at a craft store), but a disposable food container or a small storage bin would work just as well. Just make sure you're willing to make a hole in whatever you decide to use.

The size of the Styrofoam blocks that you need depends on the size and weight of the plastic container you use. The pictured boat uses two 2″ × 12″ × 4″ Styrofoam blocks. You can likely get away with smaller ones, but Styrofoam is cheap, and it is safer to use a little more, rather than use too little and have your boat sink as a result.

The silicone/glue will be used to seal up any holes that you need to make in your boat during its construction, so it's very important that this is thick enough to make a good seal, and entirely waterproof. Clear silicone is designed for creating waterproof seals, but something like E6000 QuickHold Contact Adhesive also works well.

Bill of Materials

Table 3-1 lists the materials you'll need to build your boat.

Table 3-1 *Bill of materials*

Count	Part	Source	Estimated price
1	Tamiya submarine motor 70153	Tamiya or Amazon	$10
1	Spark Core	*http://spark.io*	$40
1	Sparkfun motor driver 1A dual TB6612FNG	*http://sparkfun.com/products/9457*	$9
1	Row of break away headers (you need 16 pins)	*http://sparkfun.com/products/116*	$1.50
1	Standard servo	Electronics retailer	$10
1	0.1uF Ceramic capacitor	Electronics retailer	< $1
1	Half-size breadboard	Electronics retailer	$5
1	4 AA battery holder	Electronics retailer	$2
4	AA batteries (spares are recommended)	You probably already have these	$3
1	Pack of at least 15 male to male jumper wires (you'll need at least 2 long ones)	Electronics retailer	$2
1	Tube of Silicone or a goopy waterproof craft glue	Local craft or hardware store	$5
1	Sealable plastic container that comfortably fits your breadboard and battery holder	Multiple options	$5
2	Styrofoam blocks	Local craft store	$4
5+	Popsicle sticks	Local craft store	$2

Tools

You will also need to use the following tools to put your boat together:

- USB Microcable for setting up your Spark Core
- Wire cutters/strippers
- Soldering iron
- Spool of solder
- Drill with approximately a ⅛″ drill bit
- Hot glue gun with glue
- Bathtub or kiddie pool

The Submarine Motor Pod

When building a boat, you need to be very careful to keep the electronics out of the water and keep them from shorting out. Otherwise, it's a quick end to your project. But if you need to keep your electronics dry, how will you propel the boat through the water?

Say hello to the *Tamiya Submarine Motor*! This is a wonderful waterproof pod, designed to keep the motor safe and dry inside. There is a catch, however: normally, the pod is attached via a suction cup, and can't be externally controlled. You'll need to modify this so that you can drive it from the Spark Core.

Because the Tamiya uses a simple, bidirectional motor, you can drive it forward and backward with a *motor driver*. Some motor drivers, including the one you'll be using, will also let you control the speed. You'll need to modify the motor and the pod, however, to add wires that you can hook up to the other electronics.

Why Use a Motor Driver?

You can simply connect most I/O components, from LEDs to servos, to the appropriate pins on the microcontroller and reasonably expect them to work properly (with the possible need of capacitors or resistors). This isn't the case for motors.

The problem you face is that the motor draws too much current. In fact, it draws more current than an Arduino or Spark can provide, and more than can even safely travel through the chip! This means that if you plug a motor into your microcontroller, it may work if the motor is very small, but you run a huge chance of frying everything beyond repair.

The way to solve this problem is by using a motor driver or a type of chip called an H-Bridge. These allow you to easily bring an external power source to your motor, and the chip is designed to be able to handle the current.

Using a driver or H-Bridge also brings additional benefits. Some motor drivers, such as the SparkFun driver we're using to build this boat (Figure 3-2), also incorporate some extra components that allow you to control the speed at which the motor spins.

Figure 3-2 *SparkFun motor driver*

Motor Pod Components

The Tamiya is normally used by inserting a battery inside the pod, but we're going to modify it to use an external power source, which will allow you to turn it on and off from your computer. As such, you don't need all of the compo-

Tamiya variants

These modifications are based on the yellow Tamiya motor. If you have purchased the red Tamiya Mini, the modifications will follow the same concepts, although the guts of the pods vary a bit and the red version has a smaller chassis to work in.

nents included in the sub kit. You'll need the external casing for the pod, some of the blue plastic punch-out components, the black rubber ring, the small tube of grease, and the motor.

You can discard most of the blue plastic components that come with the pod. You will later fashion your own rudder, because the one included in the submarine pod is too small to steer a whole boat. However, you do need two of the included pieces. One piece is the propeller, the other is a small ring that is flat on one end and clips over the motor pod, pictured in Figure 3-3.

Figure 3-3 *Plastic ring that comes with the Tamiya kit*

Punch out the propeller and clip and set them aside until you're done with the soldering.

Modifying the Motor

In order to run an electrical current to the motor to control it, you will need to solder two wires onto the motor. You'll then run these wires to the Spark Core inside the boat. There's not a lot of room inside the motor pod, so you need to make sure that the wires you're soldering onto the motor are flexible and thin. As it turns out, standard jumper wires are perfect for this. Here are the steps you should follow:

1. Take your two longest jumper wires. If your jumper wires have a thicker, knobby end, clip it off with your wire cutters. Then strip the rubber insulation off one end of each wire, as shown in Figure 3-4. Make sure not to cut through the metal wires. You simply want to cut away the rubber to expose the wire underneath.
2. Take a look at the motor that came with your Tamiya. You'll see that there are two little metal loops sticking out on one side. These are the positive and negative terminals that are used to power the motor, and are what you will connect your exposed jumper wires to.

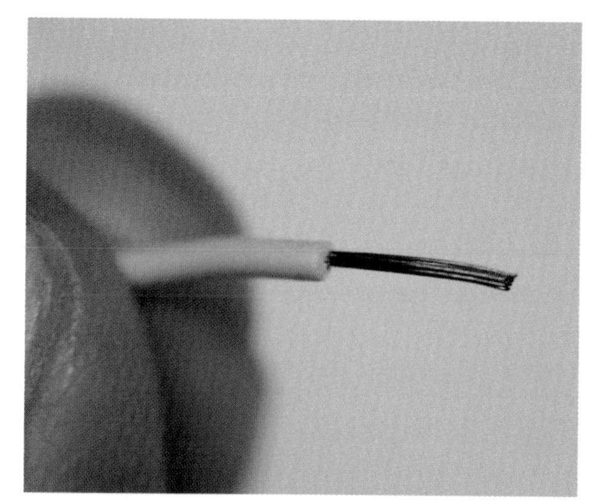

Figure 3-4 *Stripped jumper wire*

3. Loop the exposed metal of the jumper wires through the metal loops on the motor to help secure them in place.
4. You want to position the wires so that the long ends are against the body of the motor, pointing backward toward the spindle/propeller end of the motor. If you want, you can tape the wires to the body of the motor to help keep them in place while you solder.
5. Solder the jumper wires to the terminals on the motor, as shown in Figure 3-5, keeping the jumper wires toward the outer edges of the loops. It doesn't have to be perfect, but later on you'll need to be able to fit the blue plastic clip that you set aside earlier on the end of the motor.

Figure 3-5 *Soldering the wires to the motor*

Make sure that the exposed metal at the end of the jumper cables does not touch the metal body of the motor. This could cause a short circuit and prevent your motor from working. If you're worried this may happen, you can wrap some tape around the motor where there may be contact.

Testing the Motor

At this point, it's a good idea to check your motor and make sure it's soldered up correctly. To do this, you can simply take the two wires you soldered onto the motor and touch each to opposite ends of one of your AA batteries. If the motor starts spinning, you're well on your way to making a boat! If it doesn't, take a look at your solder joints to make sure a solid connection is being made, and that nothing is shorting out. Try using fresh batteries.

Finishing the Motor

Once you complete the soldering and confirm that the motor is working, we need to take a couple more steps before it's ready to go inside the motor pod (you'll need the blue plastic ring that you set aside earlier, the black rubber ring, and the small tube of grease):

1. Place the blue plastic ring over the white end of the motor, making sure the clip is on the side with the wires. The clip should fit snugly over the little protrusion on the side of the motor. See Figure 3-6.

Figure 3-6 *Clip the ring onto the motor*

Normally, in the sub motor kit, this plastic piece would hold the leads in place that connect the motor to the battery. However, for our purposes, it will help ensure the motor is secured at the back of the pod and remains in contact with the propeller while closed.

2. Apply grease to the flat end of the rubber ring. This will help make a seal to prevent your motor from flooding. (Don't use all the grease now; you'll need more later.)
3. Place the rubber ring over the spindle end of the motor, with the flat end of the ring against the body of the motor, as shown in Figure 3-7.

   You want the grease to be filling the space between the body of the motor and the rubber ring. If there are any noticeable gaps, you can add some more grease now.

Figure 3-7 *Insulating rubber ring*

Inserting the Motor

Now it's time to insert the motor into the pod to keep it dry!

1. Take the back half of the Tamiya pod. It should have a hole in the back end of it.

   Look inside and find the plastic spacer jutting a little toward the center of the pod's cavity (Figure 3-8). When the motor is inserted in the pod, this extra bit of plastic could get in the way of your wires, so take note of its position.

Figure 3-8 *Motor pod interior*

2. Double back the jumper wires, so that they are pointing away from the motor's spindle. Hold them against the body of the motor. The two wires should be close together, toward the center of the motor, as shown in Figure 3-9.

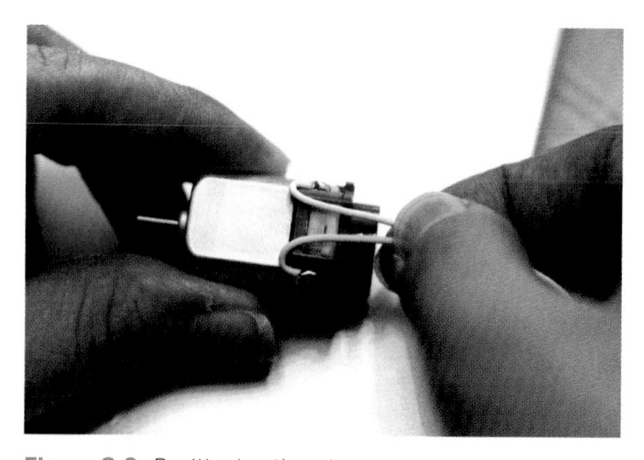

Figure 3-9 *Positioning the wires*

3. Holding the wires in place, insert the motor into the back half of the motor

pod, spindle first, as shown in Figure 3-10.

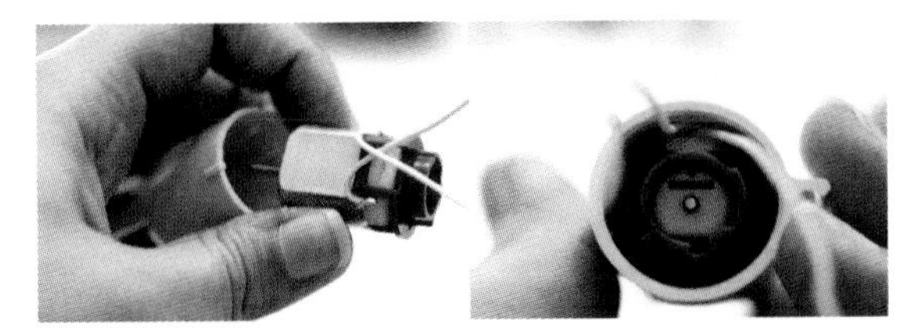

Figure 3-10 *Inserting the motor*

As you're inserting the motor, make sure that the wires are protruding from the pod. There is a small gap in the blue plastic ring that they can fit through.

4. Push the motor into the pod until it fits in place snugly and the spindle protrudes from the back end of the pod.
5. Take the plastic propeller that you set aside earlier and push it on to the motor's spindle, as shown in Figure 3-11. Once the propeller is in place, the back half of your motor is good to go!

Figure 3-11 *Propeller*

Drilling into the Motor Pod

Of course, you still need to close up your motor pod, and make sure there's a place for the wires to exit so that you can run them to your Spark. To do this, you need to drill into the body of the sub. There are a couple of things you should keep in mind when deciding where to drill the hole.

There's a lot of space to work with in the front of the pod, but the casing tapers toward the nose. The further into the nose you try to reach your wires, the harder it will be to thread the wires through the hole you'll drill. Making a hole closer to the nose would also waste valuable length of wire!

If you look inside the pod, there's a track that runs along one side. This would usually be used to secure part of the battery holder. You probably don't want to drill through this because it makes the wall of the pod thicker. Finally, the wires should exit the pod on the side that will be facing the boat. There is an external protrusion on the outside of the motor pod, make sure this won't interfere with your placement either. Once you're ready, follow these steps:

1. Line up the external nubs on the front and back halves of the sub, as shown in Figure 3-12. You will use these to help guide where you'll drill the hole.

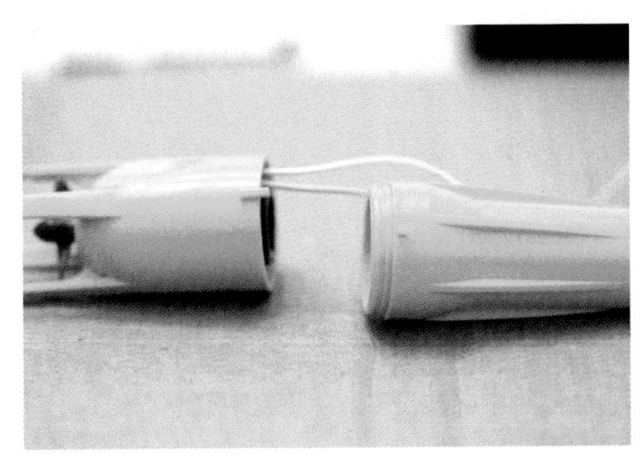

Figure 3-12 *Submarine Pod pre-assembly*

   Mark where on the front half of the sub you want the jumper wires to exit.

2. Use a 1/8" drill bit to make a hole at the location you marked on the pod casing. You want the hole to be as small as possible while still being able to thread both wires through it.

Check and see if the jumper wires emerging from the rear end of the motor pod can fit through the hole you drilled. If not, use the drill to enlarge the hole until both wires fit through.

3. Once the hole is large enough for the wires to fit, thread the wires through one at a time, from the inside of the motor pod. See Figure 3-13.

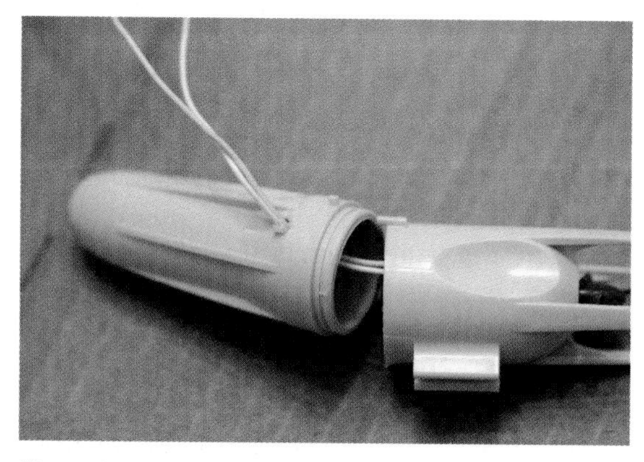

Figure 3-13 *Tamiya pod after wiring modifications*

Closing the Motor Pod

Now that your motor is wired up and the wires are accessible outside the pod, it's time to close the pod and seal the hole in the hull of the sub:

1. Use the remainder of the grease that came with the motor kit around the seam where the two halves of the pod meet. The grease will make a better seal.
2. Snap the two halves of the pod together and wipe off any residue that leaks out from the seam.
3. Pull the wires so they are taut, taking care to not pull them loose. Take the silicone or glue sealant, and place it around and over the hole you drilled in the pod, making sure there are no gaps.
4. Set the pod aside and let it dry. If you have a quick-dry formula, it should only take a couple of minutes before it's dry enough to move around.

Although most of these silicones and glues take up to a day to completely cure, if you're building a boat in a limited time frame you probably don't have to wait quite so long. Most of these will be dry and waterproof enough for you to set sail in a couple of hours. This is one of the reasons why it can be a good idea to work on the physical build of your boat first.

Waterproofing Your Wires

Maybe you've soldered your wires to the motor, and now you're realizing they're too short. Or maybe later in your project you'll decide you want to position the motor pod further away from the body of the boat and need more length. Either way, if you need to extend your wires, you need to make sure you waterproof them, too.

You'll need to take additional steps to keep your soldering water-tight. Shrink wrap tubing can help with this. You can buy heat guns to shrink the tubing, but some shrink wrap can be heated up simply with a good hair dryer. Make sure you buy shrink wrap that's compatible with the tools you have available. Then:

1. Before you solder the two wires together, slide a piece of tubing of an appropriate diameter onto one of the wires.
2. Solder the two wires together.
3. Position the shrink wrap tubing fully over where you joined the two wires.
4. Heat the tubing until it shrinks enough to form a good seal around the wire.
5. If you're concerned that the shrink-wrap didn't form a good enough seal, you can enforce the ends with some of your silicone or waterproof glue.

Your wire should now be able to withstand submersion.

Setting Up Your Spark

Now that you have completed your pod, you'll have to claim your Spark so that you can begin wiring up the hardware and controlling it from your computer. Check out "Spark WiFi Development Kit" if you haven't done this yet; it will walk you through the process.

Figure 3-14 *Spark Core*

Some Spark Core chips have a Chip Antenna, while others have a uFL Connector. The Chip Antenna version includes an onboard antenna. If your Spark has a uFL Connector, you'll need to connect an external antenna—either a flex antenna or a "duck" antenna. Make sure your antenna is in place before trying to connect your Spark to the WiFi!

Although you can claim your Spark over WiFi using a mobile app, the process can sometimes go smoother if you connect your Spark via USB and use the Spark CLI command-line tool (*http://docs.spark.io/cli/*), particularly if there are a lot of WiFi devices or any other Sparks around. You don't want to accidentally claim someone else's device.

Because you'll be configuring your Spark for your current WiFi network, remember that if you take your boat to another location with a different WiFi network, you'll have to reconfigure it and go through the `spark setup` again! Keep this in mind when assembling your boat. You don't want to have to disassemble the whole thing just to set up your Spark's WiFi credentials.

During setup, if at any point your Spark begins flashing unexpected colors and you want to know what it means, you can find out in the documentation (*http://bit.ly/19LX8iW*).

Before getting to the code for your boat, you will also have to replace the Spark's stock firmware with VoodooSpark (*http://bit.ly/19LX8zv*). This is also outlined in "Spark WiFi Development Kit".

Testing the Spark

Before continuing, connect the Spark to the battery pack and test it to make sure it is working properly and can receive commands from your computer:

1. Plug your Spark into the breadboard, making sure that it spans the divider running down the middle of the breadboard.
2. Connect the red wire from the battery pack to the pin on the Spark labeled `VIN`.
3. Connect the black wire from the battery pack to `GND` on the Spark. If you hold the Spark so that the USB port is facing up, `VIN` and `GND` will both be on the top left of the board. Figure 3-15 shows how it should be wired.
4. Put four AA batteries in your battery holder. The Spark will turn on, flash green to signal that it is connecting to the WiFi, and then "breathe" cyan (a slow fading in and out of the light as opposed to a steady blinking) once it connects.

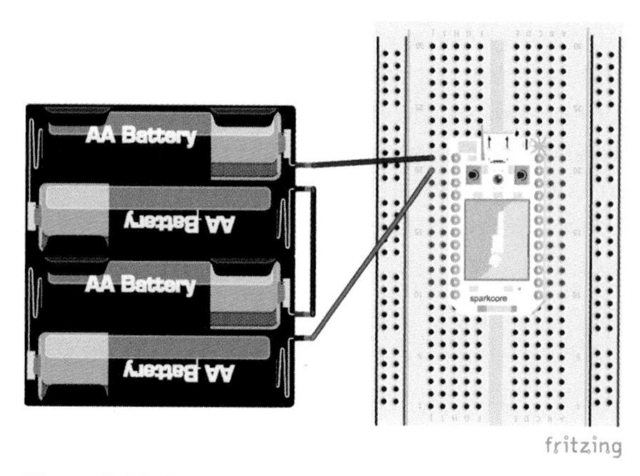

Figure 3-15 *Wiring for Spark test*

Your First Spark Program

Now you're going to create a simple test program to make sure that your Spark can properly receive commands over WiFi. This is the "Hello, world" of electronics, blinking an LED, shown in Figure 3-16. The Spark has an onboard LED, which shares a connection with the pin `D7`, so you will create a simple program to blink this light.

Figure 3-16 *Running the Spark test*

At this point in the book, you've probably gone through the basic Johnny-Five setup. If not, it may be useful for you to take a look at "Installing Johnny-Five". Then follow these steps:

1. Before you start coding, you'll need to create a directory for your project and change directory (`cd`) into it.

   Once in your project directory, you can begin your project by creating a *package.json* file. You can do this with the command `npm init`. This will walk you through the steps of creating your project's *package.json*, which is where a list of the dependencies for your code will be stored.

2. Because you're using a Spark instead of an Arduino for this project, you'll need to use the *Spark-IO* module as a plug-in for Johnny-Five. In your project directory, install Johnny-Five and Spark-IO. Add the `--save` flag to automatically add these modules to your *package.json* file:

```
npm install --save johnny-five
npm install --save spark-io
```

Now make a new file called *test.js*. This is where you'll write your code to control the Spark, shown in Example 3-1.

Example 3-1 *Spark test script for Johnny-Five*

```
var five = require("johnny-five"); ❶
var Spark = require("spark-io");

var board = new five.Board({ ❷
  io: new Spark({
    token: <your_access_token>, ❸
    deviceId: <spark_device_id>
  })
});

board.on("ready", function() {
  var led = new five.Led("D7");
  led.blink();
});
```

❶ In this file, you should start by requiring the Johnny-Five and Spark-IO modules, and setting them equal to variables. This will allow you to access the Johnny-Five and Spark-IO modules in this file.

❷ Initialize a new Johnny-Five Board. When you do this, you can pass in a JavaScript object as an argument. Because we're using a Spark, and will be using the input/output of the Spark instead of an Arduino, we will specify the `io` and set it equal to a `new Spark()`.

❸ Now you're making sure your program knows to look for a Spark, but it still does not know how to identify your particular Spark. In order to do this, you will need to pass a JavaScript object as an argument to your new Spark object. You will need to specify the `token` and `deviceId` that you noted while configuring your Spark.

It's a good idea to save your access token and device ID as environment variables to keep them out of your source code, as explained in "Spark WiFi Development Kit".

Once you save this code in your *test.js* file, go ahead and run it from the command line with the command `node test.js`. The small, blue LED next to the star symbol (shown in Figure 3-16) printed on the Spark board should begin blinking steadily. Good job—your Spark is all set up! Remember, if you run into any problems, you can check out the Spark troubleshooting guide (*http://bit.ly/1bQQucL*).

Be careful: the chip on the Spark can get pretty hot once it's been running for even a few minutes!

Unplug your battery pack for now, because you'll wire it up a bit differently for your actual boat build.

Soldering the Motor Driver

Now, before moving on to building the boat, there is one more bit of soldering you need to do. If you take a look at the SparkFun motor driver, you'll notice that there are no headers attached. You won't be able to plug it into the breadboard until you solder some on, so let's take care of this now:

1. The motor driver has two rows of holes. Hold the driver with the chip facing up.

   Take your strip of breakaway headers and snap off two rows of eight pins each. Position the two rows of headers underneath each of the rows of holes in the driver. The short ends of the header pins should be coming up through the holes in the driver. When both sides of headers are in position, the motor driver should be able to stand on its own like a table.

2. Solder the headers in place, from the top side of the motor driver, as shown in Figure 3-17. Make sure that you don't accidentally connect two pins together with the solder. This would short-circuit your driver and cause it to malfunction.

Figure 3-17 *The soldered motor driver*

Wiring Up Your Boat

You're going to have three main components inside the boat. First, you have the motor driver. Next, you have the Spark Core. And finally, you have the battery holder. All three of these will be plugged into your breadboard and wired up to communicate with each other:

1. Plug your Spark and motor driver into your breadboard. They should fit next to each other comfortably. Make sure the Spark and the driver both span the divide in the center of the breadboard, as shown in Figure 3-18.

Figure 3-18 *The Spark and motor driver*

You should position your Spark so that the USB port is facing the outer edge of the breadboard, just in case you need to access it later on.

2. Connect the battery holder to the breadboard, as shown in Figure 3-19. Plug the red wire of the battery holder into one of the side columns of the breadboard with a + sign on it. This column will be your power. Now plug the black wire into the column of the breadboard with a - sign on it, right next to it. This will be your ground. Whenever you need access to power or ground for your Spark, driver, or motor, we can connect to these strips on the breadboard.

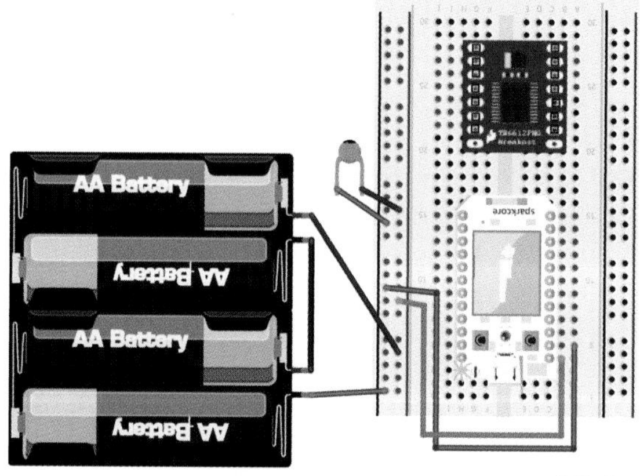

Figure 3-19 *Wiring the power for the boat*

3. Add your ceramic capacitor to the breadboard. You want to plug one end into the power, and the other end into the ground. With a ceramic capacitor, you can plug either end into power and the other end into ground, but for some other types of capacitors the direction matters. You can think of the capacitor as a type of reserve power source, that will help fill in gaps if the hardware needs more power than the battery can provide at once.

The Spark can be powered by an external power supply ranging from 3.6–6.0 Volts. Each AA battery provides 1.5V, so the four AA battery pack is perfect. If you use rechargeable batteries, be aware that each battery provides only 1.2V when fully charged.

4. Once your capacitor is in place, you'll connect the power to your Spark. This time, instead of running the power and ground directly from the battery, you'll

use jumper wires to connect the power and ground columns of your breadboard to the VIN and GND pins on the Spark. The Fritzing diagram (*http://fritzing.org*) in Figure 3-19 shows this.

Powering the Motor Driver

Now it's time to wire up the motor driver. The driver board does have labels printed for each of its pins, but unfortunately they're printed on the underside of the board. If you're looking at it from the top, the side with the chip on it, the pins are set up like in Figure 3-20.

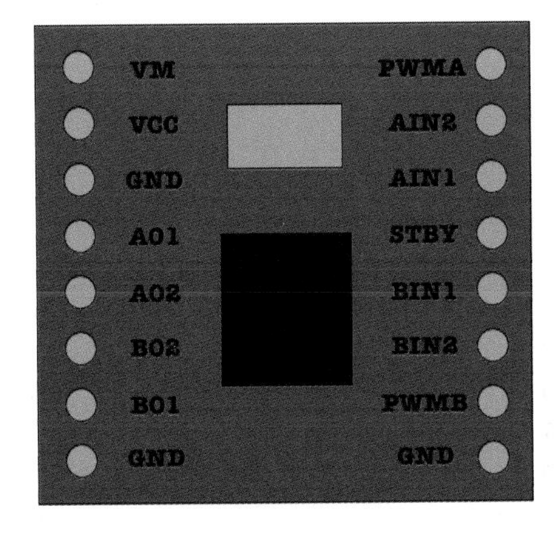

Figure 3-20 *Motor driver diagram*

In order to power your motor driver, you will need to provide two sources—one from the battery and one from the Spark:

1. Find the VM pin on the motor driver. Use a jumper wire to connect the power from the battery to this pin. Next, find the GND pin, and connect it to the battery's ground the same way.
2. In between the VM and GND pins on the motor driver there is another pin labeled VCC. You will need to connect another power source to this, but this pin only accepts between 2.7V and 5.5V. The battery pack provides more than this, so where will you get your power from?

   The solution is the Spark. If you take a look at the Spark, there is a pin labeled 3V3 which can supply 3.3V! You can simply run a jumper wire from the Spark's 3V3 pin to the Driver's VCC pin.

Spark has two 3.3V pins. One is labeled 3V3, and the other is labeled 3V3. (The asterisk is printed very small, so it looks more like a dot.) Make sure to use the one without the star when wiring up your project! The asterisk indicates that this is a low-noise regulated power rail, and it may not be able to provide enough power.*

Connecting the Spark and the Motor Driver

The motor driver we're using supports two motors, but we will use one for this boat. The three pins on the motor driver that we care about for our motor input are PWMA, AIN1, and AIN2. To connect the Spark and the motor driver, follow these steps:

1. Use jumper wires to connect PWMA on the motor driver to A0 on the Spark. This is one of the Spark's PWM pins. PWM stands for pulse-width modulation. This will allow you to adjust the speed at which the motor spins.
2. Connect AIN1 and AIN2 on the motor driver to D0 and D1. These two input pins will each correspond with the motor spinning in a different direction. Figure 3-21 shows the Fritzing diagram for the wiring.

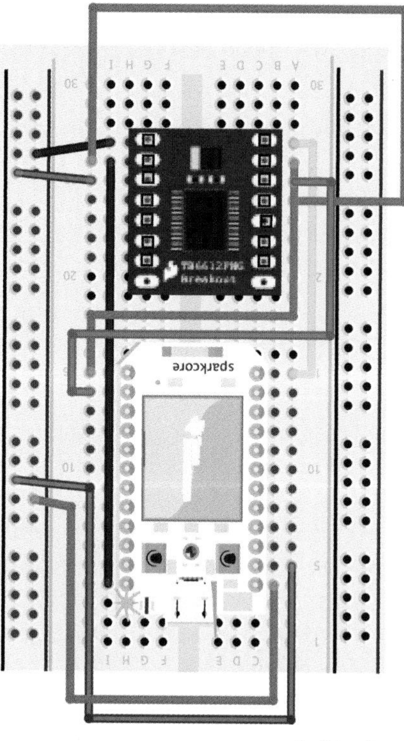

Figure 3-21 *Wiring the motor driver*

Pulse-Width Modulation Pins

Not all pins are PWM pins. On the Spark the pulse-width pins are A0, A1, A4, A5, A6, A7, D0, and D1.

What makes a PWM pin special? Most pins can only be turned on or off. PWM pins are different because they can be set to turn on and off incredibly fast. You essentially pick the percentage of the time the pin is on, and this allows additional levels of control over components.

An LED is a good example of this. If it is connected to a non-PWM pin, it can only be turned on or off. Plug it into a PWM pin and that all changes. Now the light has the ability to pulse faster than you can see, like flickering a light on and off. This gives the impression of it getting brighter or dimmer, based on the percentage of time the pin is on.

Connecting the Motor

Now that the motor driver is receiving power from the battery and is wired to receive commands from the Spark, it's time to connect the motor that we've sealed inside the Tamiya pod:

1. Take one of the wires emerging from the motor pod and connect it to `A1` on the motor driver. Connect the other motor wire to `A2`. Keep in mind that because this is a bidirectional motor, which wire is connected to which pin does not matter here.

In this example, you're using one motor pod to propel the boat, and you will later add in a servo to act as a rudder. However, if you wanted to, you could add in a second motor pod. Luckily, the motor driver can control up to two motors! If you were to do this, your second motor would be plugged into `B01` *and* `B01` *on the driver board. The corresponding pins would be* `BIN1` *and* `BIN2` *to control the direction in which the motor spins, and* `PWMB` *to control the motor's speed.*

You can use two motors simply to get more speed, or you could position one on each side of the boat and steer by alternating the speed and direction of the two motors.

2. There's one more bit of wiring you need to do for your motor to work properly. You'll notice that the motor driver has a pin labeled `STBY`. This pin will prevent your motor from turning on unless power is given to it. We don't really need this for our boat to function properly, so you'll effectively turn this

feature off by connecting this to the Spark's `3V3` pin so that it always receives voltage. Figure 3-22 shows the final wiring for the Spark and motor.

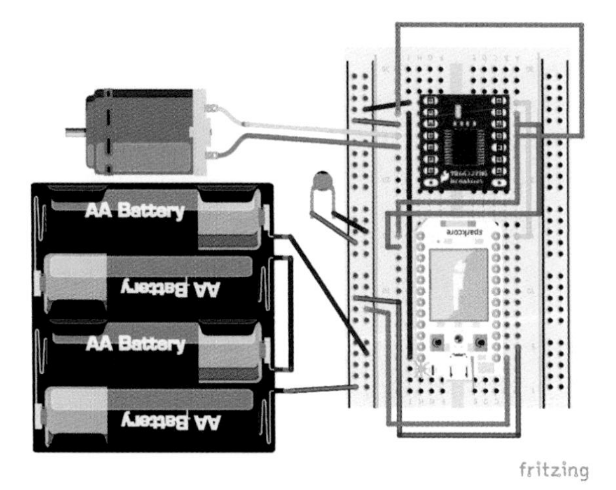

Figure 3-22 *Complete wiring of Spark and motor*

Controlling Your Motor: The Code

Now that the basic motor of your boat is now complete, let's start coding.

All source code for the examples in this book can be found on GitHub (*http://bit.ly/19LX9n3*).

1. Create a new file in your project called *boat.js*. Set up this file similar to the *test.js* you created earlier, requiring Johnny-Five and Spark-IO. This time, though, in your ready event, you will need to instantiate a new motor object instead of a new LED.

2. When you create a new motor, pass a JavaScript object as an argument. This object will specify the three Spark pins that we have connected to the motor driver. You need to specify which is the pulse-width module (`pwm`), which is the forward direction (`dir`), and which is the reverse direction (`cdir`). If you have been following along, these will be `A0`, `D0`, and `D1`, respectively:

```
var motor = new five.Motor({
  pins: {
    pwm: "A0",
    dir: "D0",
    cdir: "D1"
  }
});
```

3. In order to more easily test your motor and control it from the terminal, you can also add a REPL to your *boat.js* file:

```
this.repl.inject({
  m: motor
});
```

   Remember, instantiating the motor and adding it to the REPL should all take place inside your ready event.

4. Connect your battery pack and insert the batteries so that you can test the motor.

   Wait for the Spark to connect to the Wi-Fi and breathe cyan. Then run your program with `node boat.js`.

   Once your terminal prints out `Repl Initialized`, your Spark is ready to receive commands. If you've wired everything up properly, you should be able to control the direction and speed of your motor!

5. In your terminal, you can now run `m.forward()` and `m.reverse()` to make the motor spin in one direction or the other. `m.stop()` will stop the motor. But don't forget, with this motor driver you can also control the speed of the motor!

   To do this, you can pass a number, from 0 to 255, as an argument to the forward and reverse functions. 0 is equivalent to the stop function, and 255 is the top speed!

As you're testing your motor, you can hold your hand behind the motor pod and feel for air to see which direction the motor is spinning. If your forward function is going to make your boat spin in reverse, there are multiple ways you can fix this. The easiest way is to simply switch the Spark's `dir` and `cdir` pins in your code. (Remember, we're using `D0` and `D1` for these.) You could instead reverse the two wires plugging into the motor driver's `A01` and `A02` pins if you wanted a physical solution.

Adding Keypress Events

Of course, while you could just drive your boat from a REPL in the terminal, it would be a lot friendlier to be able to drive it with directional keys. To do this, include the Node `keypress` module:

```
npm install --save keypress
```

The keypress module will let us listen for keypress events, and you can specify which keys to listen for. It also provides convenient names for your keys, so you can listen for a keypress on the up key by looking for `key.name === "up"`:

1. Require "keypress" in your code, and make sure that `process.stdin` will emit keypress events.
2. On keypress events, specify what action you want taken if different keys are pressed. In this case, "up" should call `motor.forward(255)` and "down" should call `motor.reverse(255)`, as shown in Example 3-2.

Example 3-2 *Node keypress events*

```
var keypress = require("keypress");

board.on("ready", function() {
  // make process.stdin begin
  // emitting "keypress" events
  keypress(process.stdin);

  // listen for the "keypress" event
  process.stdin.on("keypress", function (ch, key) {
    if (key.name === "up") {
      motor.forward(255);
    }
  });
});
```

3. The keypress module does not detect keyup events, so with this code, your boat will never be able to stop. You'll need to add in another key as a brake. The space bar is a good choice:

```
if (key.name === "space") {
  motor.stop();
}
```

Storing Keypress State

One of the problems we have with using the keypress module is that if you hold down a key, the function tied to that key will continually be called until the key is released. This can overwhelm your Spark and crash it. In order to avoid this, you can store the state of the keypresses in an object and only call the function when the key is first pressed:

1. Create an object and set it equal to the variable `state`. The keys will be `up` and `down`, and you should set the values of both to `false`.
2. When you detect the keypress event in your code, adjust your functions so `motor.forward()` and `motor.reverse()` are only called if the related up or down state is false. Likewise, you will only want `motor.stop()` to be called when either the up or down state is true.
3. When `motor.forward()`, `motor.reverse()`, or `motor.stop()` is called, you want to make sure you adjust the values of the keypress states accordingly.

So now the simple going forward function will look more like Example 3-3.

Example 3-3 *Controlling the motor through keyboard events*

```
var keypress = require("keypress");

var state = {
  up: false,
  down: false
};

board.on("ready", function() {
  keypress(process.stdin);

  var motor = new five.Motor({
    pins: {
      pwm: "A0",
      dir: "D1",
      cdir: "D0"
    }
  });

  process.stdin.on("keypress", function (ch, key) {
    if (key.name === "up" && !state.up) {
      state.down = false;
      state.up = true;
      return motor.forward(255);
    }
    if (key.name === "down" && !state.down) {
      state.up = false;
      state.down = true;
      return motor.reverse(255);
    }
    if (key.name === 'space' && (state.up || state.down)) {
      state.down = false;
      state.up = false;
      return motor.stop();
    }
```

```
    });
  });
```

Floating the Boat

Now that you have a way to propel your boat and have finished wiring the internal components, you can finish up the physical build:

1. Take the two pieces of Styrofoam and use the hot glue gun to attach one to each side of the boat, as shown in Figure 3-23. While you're positioning them, remember that you want the motor pod to be fully submerged, but need to make sure that the top of your boat is above water.

Figure 3-23 *Assembling the boat hull*

   Any electronics, including the servo that you will later rig up to the top of the boat, must be kept out of the water or they will short-circuit and your boat will not work. Keep this in mind when sticking the foam in place. You also want to make sure that the positioning of the foam doesn't prevent your lid from closing.

2. Place your breadboard and battery pack inside the container serving as your boat hull, and decide where you want to position your components. Once you have decided, mark the location where you want the wires from the motor pod to enter your hull.

At this point, you may want to fill the container with your components, or items of equivalent weight, and check its buoyancy.

   You can remove the electronic components from the boat.

3. Drill a hole in the bottom of your boat at the spot you marked. The hole doesn't necessarily have to be centered, depending on the design of your boat, but keep in mind how the placement of the hole will affect the placement of your motor pod. You do want to make sure your motor is centered, otherwise your boat may go off course.

   Just like when you drilled the hole in the motor pod, you want the hole you drill here to be as small as possible while still allowing you to thread both wires from the motor through it.

4. Disconnect the motor pod from the motor driver and feed the wires through the hole you drilled in the bottom of the boat.

5. Turn the boat upside down and position the motor pod so that it is centered and facing the correct direction. To give the motor pod stability, you want to position it against the bottom of your boat, as shown in Figure 3-24.

6. Holding the motor pod in place, seal the hole in the boat with the silicone or glue, making sure it is completely covered. You can also use the silicone or

glue to help secure the motor, although you can reinforce this later. Set the boat aside for the sealant to dry.

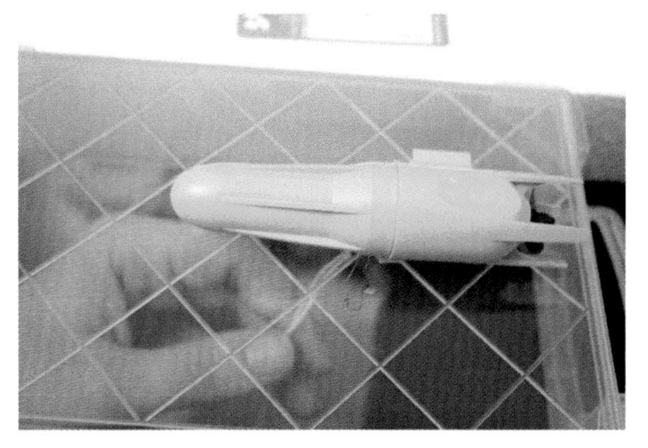

Figure 3-24 *Positioning the motor pod*

7. Once the silicone or glue has dried enough to set, flip the boat right-side up and seal the hole from the inside as well.

   Once it dries, if your motor pod is still moving around, add some hot glue between the motor pod and boat to stabilize it.

8. Put your breadboard and battery holder back inside the boat and reconnect the motor to the motor driver.

It's now time for you to test your boat and get it into water. Congratulations! But you're not done yet. Sure, you have a boat, but at this point it can only go forward and backward. Next you're going to add a rudder to steer your boat.

Make sure you're testing your boat in a small, controlled body of water, like a bathtub or kiddie pool. If it malfunctions you want to make sure that you're able to save it!

Steering with Servos

In order to control your boat's rudder, you're going to use use a servo. A servo is a type of motor that allows you to control the angle to which it can be moved, from 0° to 180°. This is perfect for moving a rudder back and forth!

We're using a standard servo for our boat. Don't confuse this with a continuous-rotation servo. Those can rotate a full 360°, and instead of controlling the angle, you control the speed at which the servo rotates.

1. Your servo should have three wires connected to it. The colors of these wires may vary between servos, but the one that is black or brown is your ground. Use a jumper wire to connect this wire into ground on your breadboard.

2. The middle wire on your servo should be red or orange. Connect this wire to your battery power, via your breadboard. Servos take a lot of power, so it's usually good to power them off of an external battery source. For this boat, the four AA batteries should be enough to power the servo, motor, and Spark, as long as your batteries are fresh.

3. The third wire on your servo is the one that will take input from the Spark. In order for the servo to work properly, it needs to be connected to a PWM pin.

   Remember, the PWM pins on the Spark are `A0`, `A1`, `A4`, `A5`, `A6`, `A7`, `D0`, and `D1`. `A0` is already being used by the motor. We will use `A4` for the servo, as shown in Figure 3-25.

Because the motor is connected to `A0`, you cannot connect the servo to `A1`. This is because `A0` and `A1` are connected to the same internal component and both will output the same frequency for a pulse width. The motor and the servo each need a different frequency to function properly, and so you need to attach the servo to `A4` to avoid the conflict. If you were building something that used two servos, connecting one to `A0` and the other to `A1` would not cause any issues.

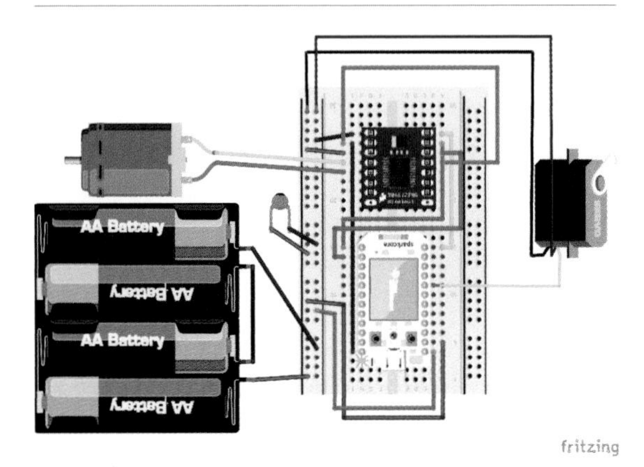

Figure 3-25 *Final wiring, with Servo*

Programming Servo Control

Now that you have a servo wired up to your boat, it's time to add it in your code, too:

1. Right after you define your motor, you'll initialize a new Servo as well, with `new five.Servo()`.

2. Pass your servo an object as an argument to set the pin to `A4`, because that's the pin you connected your servo to on the Spark.

   Set `startAt` to 90, which will make sure your servo will center itself when your program starts. You can adjust this a bit if your servo is off center.

   You can also add a range to define the minimum and maximum angles that your servo can turn to. This will make sure that your rudder does not hit into the hull of your boat, and also will allow you to define what angles you want the servo to turn to when you call the `min()` and `max()` functions. Set the range to 45–135 for now:

   ```
   var servo = new five.Servo({
     pin: "A4",
     range: [45, 135],
     startAt: 90
   });
   ```

3. In the `state` object that stores the states of the `up` and `down` keypresses, you also want to add the states for `right` and `left`. In the keypress event, you can then detect those keys and call the `servo.min()` and `servo.max()` functions to turn the servo (your soon-to-be-rudder) right and left (see Example 3-4).

Example 3-4 *Controlling the servo through keyboard events*

```
var state = {
  right: false,
  left: false
};

// ...
```

```
process.stdin.on("keypress", function (ch, key) {

  // ...

  if (key.name === "right" && !state.right) {
    state.right = true;
    state.left = false;
    return servo.max();
  }

  if (key.name === "left" && !state.left) {
    state.right = false;
    state.left = true;
    return servo.min();
  }
});
```

Now your code will allow you to steer the boat right and left on keypresses, but because the keypress module cannot detect a keyup event, your boat has no way of going straight. Ideally you won't need to press another key to have the boat start going straight again. You will use the temporal module to essentially time out the keypress, as shown in Example 3-5.

Example 3-5 *Temporal queue*

```
temporal.queue([
  {
    loop: 100,
    task: function() {
      var currentTime = Date.now();
      if (currentTime - timeOfDirection >= 500) {
        if (state.right === true || state.left === true) {
          servo.center();
          state.right = false;
          state.left = false;
        }
      }
    }
  }
]);

// ...

if (key.name === "right" && !state.right) {
  timeOfDirection = Date.now();
  // ...
```

4. Initialize a `timeOfDirection` variable, and in your keypress event, update it to `Date.now()` whenever the right or left directional key is pressed. This will allow us to see how much time has elapsed, and center the servo if the key is no longer being held down.

5. Because you will use the Temporal module, in your terminal, type `npm install --save temporal` to install the module and save it to your *package.json* file.

6. Call `temporal.queue()` in your ready event. This takes an array, and allows you to specify when tasks will run and whether they will loop. For our purpose, we will repeat only one function, and will loop over it every 100 ms.

7. In this function, compare the value of `timeOfDirection` to the current time. If more than 500 ms has elapsed, you can safely say the key is no longer being held down, and can center the servo. Make sure you reset `timeOfDirection` to the current time whenever the right or left keys are pressed.

Assembling the Rudder

Now you can control the servo! A servo, of course, is not a rudder. But we can build a rudder with Popsicle sticks and attach it to our servo.

1. Take two of your Popsicle sticks and cut them in half. Line them up so that the four halves are touching to make a square or rectangle. This will be the fin at the bottom of your rudder.

2. Glue the square or rectangle you made with the four half-Popsicle sticks to the bottom of a full-length Popsicle stick. This is now your rudder.

3. Take the end of the rudder that does not have the fin, and glue this to the rotating piece at the end of the servo. Your servo probably came with multiple attachments. You may want to use one of the longer ones so that you can position the rudder a bit further away from the boat, as shown in Figure 3-26.

4. Once your rudder is connected to the servo, you'll want to attach the servo to the back of your boat, on the outside of your boat hull. Depending on how tight the lid of your container closes, you may be able to simply close the cover over the wire leading from the boat's inner hardware to the servo. If the wire would prevent the cover from closing, however, you'll have to drill a hole in the boat lid for this as well.

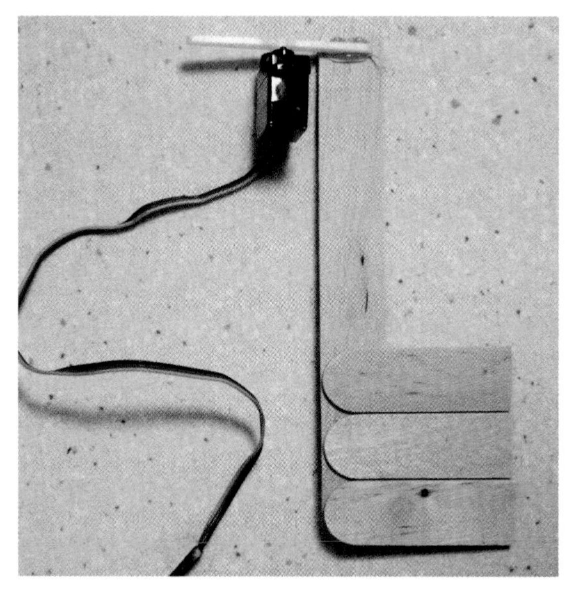

Figure 3-26 *Rudder and servo*

5. Use the hot glue gun to attach the servo to the boat. Depending on the type of container you're using and the design of the boat, you may choose to attach it to the main hull, the edge of the cover, or a clip that holds the cover in place.

   When positioning your servo, make sure that the main part of the rudder will be fully submerged in the water once the boat is afloat. Also make sure that the rudder is centered, or else it will be more difficult to control your boat.

Remember, servos are not waterproof! Keep them well out of the water, or they will stop working. Many servos have a label on the side. When gluing your servo in place, make sure you are gluing the actual servo, not just this sticker, to the boat. Otherwise, the servo may fall off and into the water when you least expect it!

Setting Sail

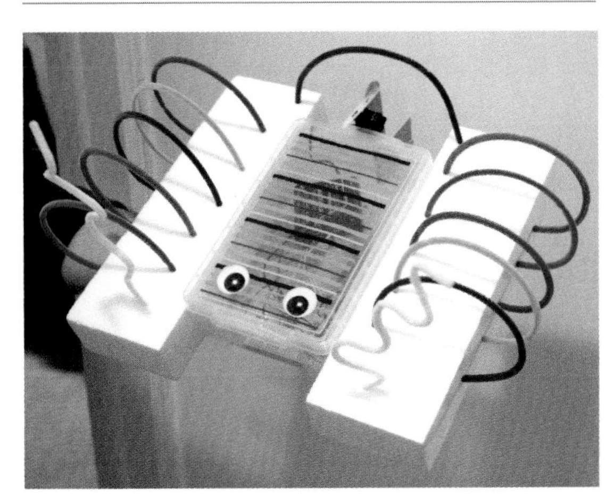

Figure 3-27 *The finished boat*

Now that you have your rudder in place, your boat is sea-worthy! Set it loose in a pool, or even just a bathtub, and watch it go!

You're also highly encouraged to decorate your boat. Whether you want it to look like a fish, a pirate ship, or something in between, the design options are limitless.

What's Next?

Now that you have a complete boat, you can work on adding more features or experimenting with other ways to control it!

For example, you could try any or all of the following:

- Adding a water sensor in the bottom of your boat to sound an early warning signal if it starts to flood.
- Steering your boat with a device that has an accelerometer, such as a Sphero.
- Include a second motor pod and steer by adjusting the speeds between the two pods.
- Programming in more fine-tuned turning.
- Add in acceleration and deceleration functions.
- Add in sonar to have a self-driving boat.
- Include LEDs that change color with the different commands you send to your boat.

There are countless possibilities, and all NodeBoats are unique! We'd love to see the creations you make. Send photos of your creations to @NodeBoats (*https://twitter.com/nodeboats*) on Twitter and have fun sailing!

piDuino5 Mobile Robot Platform

4

By Jonathan Beri

Johnny-Five was destined for greatness from the moment Rick demoed it at NodeConf 2012 (*http://youtu.be/jf-cEB3U2UQ*). Sensors, Node.js, and the greatest movie robot made for a great hardware platform, but something was absent from the demos. There was a cool tank-bot with sonar doing amazing things via a MacBook, but it was bounded by the length of the USB cable. Why wasn't this wireless? There must be a version without a cable. After tweeting at Rick and finding out that no one had tried it yet, it became a personal challenge of mine to make one of the first untethered Johnny-Five projects. The final robot used a Raspberry Pi for the main controller, an Arduino for the I/O, and of course, Johnny-Five for the software. It was cleverly called piDuino5, and it's shown in Figure 4-1.

Originally, the hardware combined both new and salvaged parts to make an untethered bot, but new parts were eventually bought online to make a clean demo. I went with cheaper and easier-to-acquire parts when possible, rather than fancy Arduino shields or high-end motors.

You may ask, "Why two boards? Couldn't you pick either the Raspberry Pi or the Arduino?" Combining a Raspberry Pi and an Arduino allows you to take advantage of each controller's extensive ecosystems of peripherals. There are all sorts of sensor shields available for the Arduino, both analog and digital, and high-performance software on the Raspberry Pi like OpenCV, but let's not get ahead of ourselves with scope-creep. We have to begin small, starting with the hardware.

Figure 4-1 *The piDuino5*

Bill of Materials

Table 4-1 lists materials used to build the original piDuino5, plus a few improvements to the design. Treat this list as a suggestion—there is more than one way to skin a Terminator, so be brave and use whatever hardware is handy. For example, the common Arduino Motor Shield is easier to assemble, but it will cost a bit more. Similarly, the Magician Chassis is the cheapest motor platform you can buy, but you could easily cannibalize an R/C car.

Table 4-1 *Bill of materials*

Count	Part	Estimated price	Part numbers
1	Raspberry Pi B/B+	$39.95	MS MKRPI5; AF 1914; SF DEV-12994
1	4GB or 8GB SD card	$11.95	MS (included w/Pi); AF 102; SF DEV-12998
1	Mini WiFi adapter	$9	AZ B003MTTJOY
1	Arduino Uno - R3	$24.95	MS MKSP99; AF 50; SF DEV-11021
1	Magician Chassis	$14.95	SF ROB-12866
1	Mini breadboard with adhesive	$3.95	MS MKKN1-B; AF 65; SF PRT-12043 through PRT-12047
1	MicroB USB breakout board	$1.95	AF 1833; SF BOB-12035
1	Break away male header pins	$1.50	AF 392; SF PRT-00116
1	SN754410 H-Bridge motor driver	$2	SF COM-00315
1	Dual-output USB battery backup	$29.95	SF PRT-11360
1	Jumper wire kit	$6.95	SF PRT-00124
1	Male to male jumper wire (30 pack)	$6-$9	MS MKSEEED3; AF 153
2	Micro USB cables	$5	AZ B00C28L5UW
1	A-Male to B-Male USB	$2.50	AZ B000FW60E8
1	Cable tie assortment	$7.50	AZ B000NQ16NG
1	Velcro strips	$6	AZ B000TGSPV6

If you plan on doing a lot of work with Raspberry Pis, pick up an FTDI Console cable. It allows you to plug in a cable into your Pi and access it via the command line from your laptop without the need to hook up a network connection or monitor, keyboard, and mouse. See Adafruit's tutorial (http://bit.ly/1wDVpr6) on accessing a Raspberry Pi from a console cable.

Tools

Most of the assembly is just screws and push pins, but some soldering is required for this project. Don't fret—a basic soldering course should be enough to build this bot:

- Soldering iron
- Solder
- Safety glasses
- Phillips screwdriver

Now let's get the Raspberry Pi out and start executing code.

Setting Up the Boards and Installing Software

In this, section you will:

- Install a Node.js build optimized for the Raspberry Pi
- Get the piDuino5 source and all its dependencies
- Connect an Arduino to the Raspberry Pi
- Blink an LED with WebSockets

Installing Node.js on the Raspberry Pi

To install Node.js on the Raspberry Pi, follow these steps:

1. Power up, log in, and make sure that your Raspberry Pi can connect to the Internet. If you need help installing Raspbian or configuring your WiFi, refer to the Appendix.
2. Install Node.js and npm. There are several ways to install Node.js onto a Raspberry Pi, but the easiest way is to download the latest version from *http://node-arm.herokuapp.com/*. This is precompiled and optimized for the Raspberry Pi's ARM architecture and includes npm:

```
wget http://node-arm.herokuapp.com/
node_latest_armhf.deb
sudo dpkg -i node_latest_armhf.deb
```

Downloading the piDuino5 Code and Dependencies

After installing Node.js, you'll need to download the piDuino5 code and dependencies. Here are the steps you'll need to follow:

1. First, you need to install the core Git package if it isn't installed already:

```
sudo apt-get install git-core
```

2. Download the piDuino5 source:

```
git clone https://github.com/
beriberikix/piDuino5.git
```

3. Change the working directory to the *piDuino5/* directory:

```
cd piDuino5
```

4. Install the dependencies and piDuino5 using npm:

```
npm install
```

This installation can take 20 minutes or more. If you want to see more information as it installs (to assure yourself it's really running), use this command to install it instead: `npm --loglevel verbose install`.

Plug in the Arduino

You'll need to upload the Standard Firmata library onto the Arduino before this step. See "Arduino" for installation instructions.

Take the A-Male to B-Male USB and plug one end into the Raspberry Pi and the other end into the Arduino (there is only one way for the cable to go).

All source code for the examples in this book can be found on GitHub (https://github.com/rwaldron/javascript-robotics).

Test Johnny-Five over WebSockets

To test Johnny-Five over WebSockets, follow these steps:

1. Switch back to the console on the Raspberry Pi and start the Johnny-Five application:

   ```
   node app.js
   ```

2. Note the local IP address in the output. It will look something like `ws://10.0.0.5:8080`.
3. You may be tempted to use cURL to connect to the socket, but cURL doesn't natively support WebSockets so you need to install a library called `wscat`:

   ```
   npm install -g wscat
   ```

4. Create a connection to the Johnny-Five application using `wscat`. Change `10.0.0.5:8080` to the address that was displayed when you ran the app:

   If you get an error installing `wscat`, try prefixing the npm command with `sudo`, or first, try installing npm-sudo-fix (https://www.npmjs.com/package/npm-sudo-fix).

   ```
   wscat -c ws://10.0.0.5:8080
   ```

5. Type `blink` into the prompt and press Enter.
6. The Arduino has an onboard LED on pin 13. It should now be blinking.

Figure 4-2 *LED on Pin 13, labeled L*

7. Type *noBlink* into the prompt and press Enter to stop the LED from blinking.

Now we're making something happen! Let's walk through the code to understand what we just did.

Walk Through app.js

The main application file, *app.js*, provides three core features:

- Uses Johnny-Five to control the speed and direction of two motors

- Responds to commands over WebSockets
- Creates a local address so commands can be sent from a device on the same network, like a web browser

Let's walk through the key parts of the code that enable those features. We'll skip superfluous code like requiring modules and error handling for the sake of simplicity.

Initializing Johnny-Five

The very first thing this program does is create an instance of a board. Johnny-Five looks for a board plugged into the computer, connects to it, and waits. Not very interesting.

```
var board = new five.Board();
```

What makes Johnny-Five interesting is the fact that it can control abstracted hardware like LEDs and motors. In order to do so, the hardware needs to be initialized with the configuration data like pin number. The configuration is typically done within the `ready` callback handler. Here's how to initialize the onboard LED on Pin 13; this is an abbreviated version of what's in the *app.js* file:

```
board.on("ready", function() {
  var led = new five.Led(13);
});
```

Led is a built-in class that is part of Johnny-Five. There are many other built-in classes. Because we're trying to control two motors, we'll need to set up two motor instances:

```
// within the "ready" handler
var left = new five.Motor({
  pins: {
    pwm: 3,
    dir: 12
  },
  invertPWM: true
});
```

Our robot is using a special type of chip called an H-Bridge. The theory behind motor control is beyond the scope of this book, but just know that this kind of chip can control the speed and direction of two independent motors.

Controlling the Hardware

With the hardware initialized, you can now make it do your bidding. Each hardware instance has its own methods that you can invoke. For example, you can make an LED blink on and off for 500ms (half second) intervals:

```
led.blink(500);
```

Motors also have their own methods, such as `forward()`, `reverse()`, and `stop()`:

```
// full speed
left.forward(255);

// half speed
left.reverse(127);

left.stop();
```

Low-Latency Control with WebSockets

WebSockets provide bidirectional communication between a web server and web browser. This is great for controlling a robot from a web browser, because we do not want any lag when issuing commands. There are many popular ways of using WebSockets but the `ws` package for Node.js is the most performant module on a Rasberry Pi at the time of writing.

The Raspberry Pi uses ARM, a different type of chip than most desktop computers and laptops. That's important, because it requires that certain libraries need to be compiled for that chip and often experience different behavior or level of performance.

Like Johnny-Five, you create an instance of a `WebSocketServer` and wait for messages once a connection is established.

```
var wss = new WebSocketServer({ port: 8080 });

wss.on("connection", function(ws) {
  ws.on("message", function(data) {
    if(data === "forward") {
      forward(255);
    }
  });
});
```

WebSockets pass simple strings around. Within the `message` callback, we look for strings that map to hardware instance functions. For example, `forward()` moves both wheels forward:

```
var forward = function(speed) {
  left.forward(speed);
  right.forward(speed);
};
```

The *app.js* example from the previous section uses code similar to the preceding examples to manage its WebSockets.

Connect from Anywhere

The current functionality allows you to send WebSocket requests from devices on the same network as the piDuino5. If you would like to control the robot from anywhere (like a smartphone), you'll need a way to expose the server to the outside world. You can do that by *proxying* requests through another server. Although I don't cover it here, you can give free services like `localtunnel` or `ngrok` a try.

Assemble the Hardware

With the controls in place, you can now start to build the platform. The fully assembled version is shown in Figure 4-3:

1. The Magician Chassis includes the basic instructions to attach the wheels, motors, and base plates. Follow it to construct the basic form.
2. Now would be a good time to affix the Velcro to the back of the battery and lower platform (Figure 4-4). Using Velcro allows you to easily access it or replace the battery without having to disassemble the entire chassis.

Figure 4-3 *piDuino5 fully assembled*

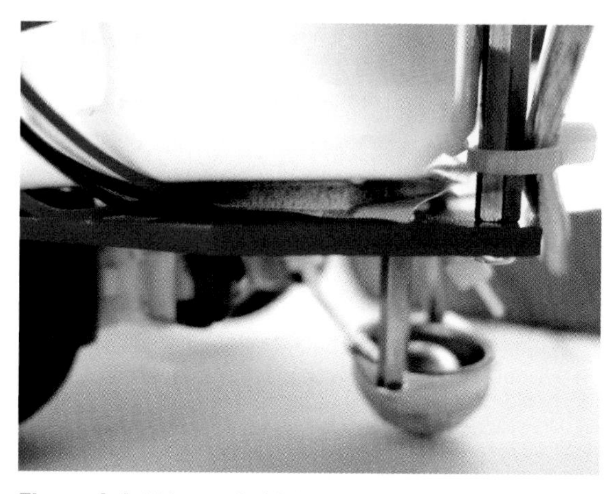

Figure 4-4 *Velcro to hold down the battery*

3. Use the extra spacers to mount the Arduino and Raspberry Pi. You can place these anywhere on the top plate, but make sure you leave room for the breadboard (next step). Use my layout as a guide.
4. Remove the adhesive backing from the mini breadboard and affix it to the top plate.
5. Plug the IC into the center of the board and start wiring up the pins to the Arduino. There are extra jumper wires in

Figure 4-5 to make the board cleaner, but these aren't necessary.

Figure 4-5 *Breadboard with H-Bridge*

Just make sure you follow the pin connections shown in Figure 4-6.

1. Plug in the motors. The motors may be reversed so try flipping their wires until you get the desired turning direction.
2. Solder the header pins to the microUSB breakout board and plug it into the breadboard. The diagram shows a battery pack in the place where the header pins should go.
3. Connect the A to B USB cable between the Arduino and the Raspberry Pi. Then connect 1 microUSB cable from the 500ma port on the battery to the Raspberry Pi and the other microUSB cable from the 1A port to the microUSB breakout on the breadboard.
4. Like we did earlier, run *app.js* on the Raspberry Pi to test the motors. Use wscat at the command line to send the message *forward*.

During testing, the robot will scurry around the table and get away from you. A simple solution is to prop up the robot so the front wheels aren't touching the surface. You may use the extra spacers that came with the kit, but you can just stand up the Magician Chassis on the flat end.

You now have a working robot that you can drive around from the command line. Pretty neat, right? We can always make it better, though. Let's add more advanced controls.

Controlling with a Smartphone

Using `wscat` from the command line is convenient but makes it hard to show off the robot to friends. A custom web app that uses buttons instead of text commands would be an upgrade. But adding support for a mobile device would be even better—and more portable. So let's make an easy-to-use HTML5 web app.

In this section, you will:

- Get the piDuino5 Web App source and all its dependencies
- Save and connect to the piDuino5's public address
- Control the robot with a virtual touch-enabled joystick

Downloading the piDuino5 Web App

To download the piDuino5 web app, follow these steps:

1. Choose where to host the web app. It can be on localhost or even a hosting provider.
2. Download the piDuino5 source:

```
git clone https://github.com/
beriberikix/piDuino5-webapp.git
```

3. Change the working directory to the `piDuino5-webapp`:

```
cd piDuino5-webapp
```

4. Install the dependencies using npm:

```
npm install
```

5. For the frontend, you'll need to install Bower and grab the dependencies:

```
npm install -g bower
bower install
```

6. Start the web app:

```
node app.js
```

Post localtunnel to the Web App

Earlier, we were outputting the local IP address to the console. Now we'll take that address and POST it to our web app. Edit the *app.js* on the Raspberry Pi and update the `webappURL` variable to the location where you're hosting the web app (it's shown as `http://10.0.0.5:3000` in this listing). This could be a host on the Internet, or an IP address (along with the port, which is the number following the colon) on your local network:

```
var five = require('johnny-five'),
    board = new five.Board(),
    PORT = 8080,
```

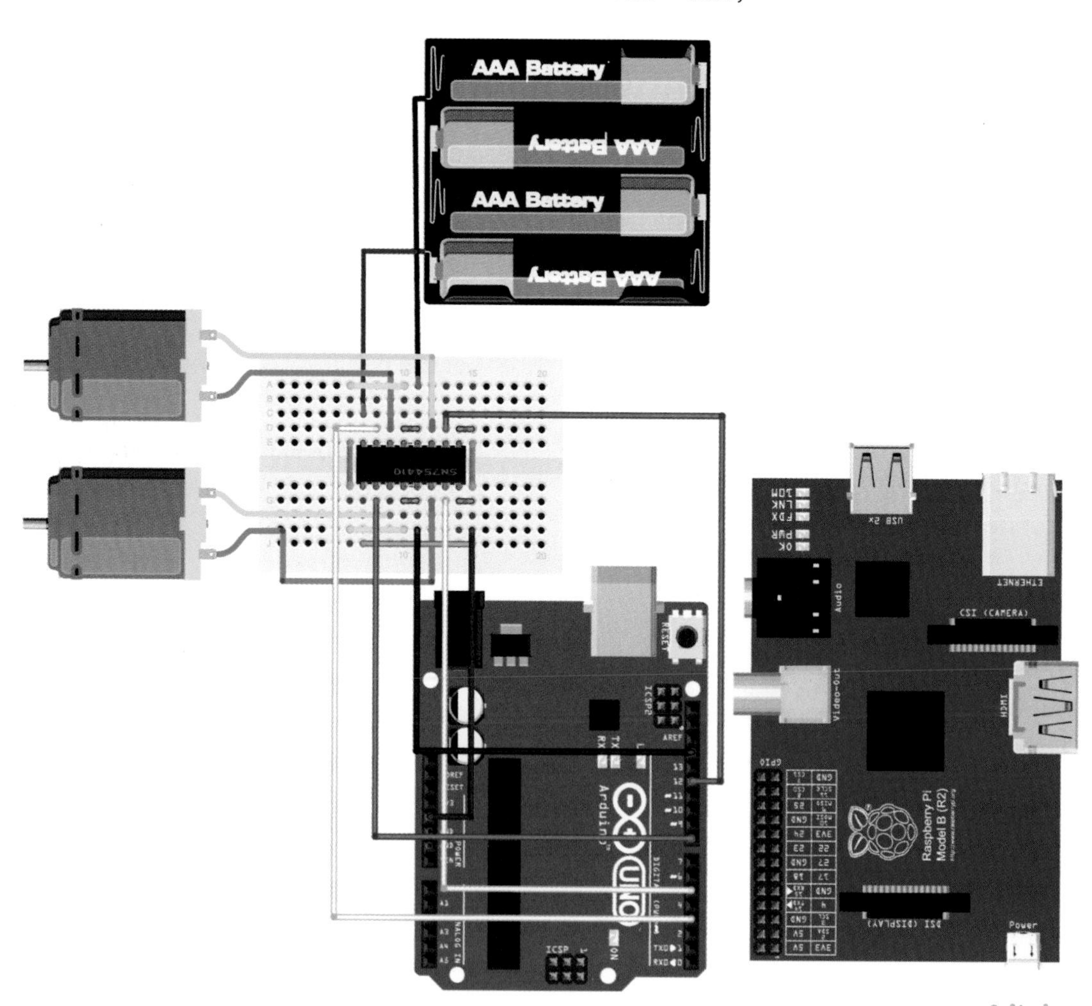

Figure 4-6 *Detailed wiring guide for the breadboard*

```
    WebSocketServer = require('ws').Server,
    request = require('request'),
    networkInterfaces = require('os').networkIn
terfaces(),
    motors = {},
    led = {},
    webappURL = 'http://10.0.0.5:3000',
    localIP;
```

Try the Web App on a Phone

1. While still on Raspberry Pi, run `node app.js`. If it's still running from before, press Control-C to stop it.
2. Navigate to the web app on a smartphone (though your laptop will work too).
3. You should see a not-so-exciting gray page (Figure 4-7) . Tap and drag upward anywhere on the screen and a virtual joystick should appear. Clicking with your mouse on your laptop should have the same effect.

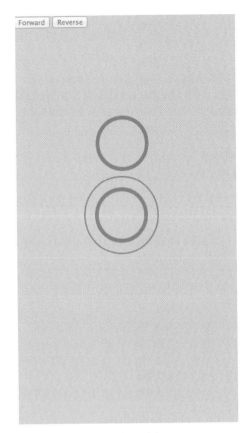

Figure 4-7 *Web app virtual joystick*

4. If all goes well, the robot should be moving forward! Try swirling the joystick around to make the robot change direction.

Don't see a joystick? Check the browser console. If you're seeing errors, it is likely that the browser hasn't made a connection with the robot. Try restarting the application on the Raspberry Pi, and then refresh the browser (in that order).

Walk Through app.js and index.html

The web app is a simple Express server that handles both the backend and frontend of the application. The core features of the web app include the following:

- An API to retrieve the address of a robot
- Code to create a websocket connection to the robot
- A joystick in JavaScript that sends commands back to the robot

Storing the localtunnel Address and Frontend

Earlier, we POSTed the local IP address to the server. The application needs a route to handle the request. In the request handler, we also parse and store the address:

```
app.post('/locate', function(req, res) {
  localip = req.param('local_ip');
});
```

Serving the UI

The application serves an EJS template from the root directory:

```
app.get('/', function(req, res) {
  res.render('index', { localip : localip });
});
```

Touchscreen Joystick

The UI portion is rather simple. There's only a single `<div>` in *views/index.html*. This is used to render a joystick using a library called *virtualjoystick.js*. Once the library loads, it finds the `<div>` and renders the entire UI. Example 4-1 is an excerpt that shows where the `<div>` resides.

Example 4-1 *Location of the <div>*

```
<div id="container"></div>

<script src="/bower_components/virtualjoystick.js/virtualjoystick.js"></script>
<script>
  var ws = new WebSocket('ws://<%= localip %>:8080');
```

One nice feature of this library is that it works with both mobile touch interfaces as well as a standard mouse.

Establishing a Connection

The application establishes a standard WebSocket connection using the lastLocation address. WebSockets are in most modern browsers, so all we need to create one is a URL. We'll pass in the URL via a template variable stored by the API:

```
var ws = new WebSocket('ws://<%= localip %>:
8080');
```

Sending Commands

The last part to cover is the most interesting bit of code—sending the commands to control the robot! You will use the joystick to control the robot so you want its direction to map to a command. When the joystick is up, the robot should go forward. When the joystick is to the left, the robot should turn left. And so on.

virtualjoystick.js has methods to check the current direction of the joystick. For example, `joystick.up()` and `joystick.left()` will return `true` when the joystick is in the obvious direction. Using a `requestAnimationFrame` loop, you can efficiently poll the joystick:

```
function step(timestamp) {
  if (joystick.up()) {
    // the joystick is up and we want
    // the robot to move forward
  }
  requestAnimationFrame(step);
```

```
}

requestAnimationFrame(step);
```

When one of the directions is `true`, all we do is send a string to the robot over WebSockets. When the robot receives the *forward* string, it responds by turning both motors on in the forward direction:

```
if (joystick.up()) {
  ws.send("forward");
}

if (joystick.left()) {
  ws.send("turnLeft");
}
```

Remember that WebSockets passes strings around. The commands chosen were somewhat arbitrary. As long as the receiving code knows how to interpret the string, anything goes. "forward" could have been replaced with "giddy-up" as long as the code on the Raspberry Pi is updated.

What's Next?

Despite all of our hard effort, the robot is kinda dumb. It doesn't know anything about it's environment. There are tons of sensors you could add to the Arduino—like an ultrasonic distance sensor to navigate around, but you can also go for something more exotic like a flame sensor and build a robotic firefighter.

The robot also doesn't use the hefty processor on the Raspbery Pi. A challenging but fun project might incorporate the Raspberry Pi camera and OpenCV (the popular image processing library) to find shapes and chase them (Robo Tom and Jerry?).

But most important—create something new! Build a bigger robot, a faster robot, or even a flying robot. Don't be afraid to dream big and create a robot as awesome as the one that inspired the framework, Johnny-Five.

Controlling a Hexapod with Johnny-Five

5

By Donovan Buck

In this chapter, you will learn how to build a simple walking hexapod—a six-legged robot. The hexapod in this chapter, shown in Figure 5-1, will have three joints in each leg. The robot will be controlled using Johnny-Five's Animation class. The Animation class is useful for scripting servos over time. It gives us a timeline, key frames, tweening, and easing functions. When you are done, you will have an excellent platform for building a more complex and interesting robot in the future.

Figure 5-1 *The finished hexapod*

Bill of Materials

The materials used in this chapter are listed in Table 5-1.

Table 5-1 *Bill of materials for robot*

Count	Part	Estimated price	Part number(s)
1	Phoenix 3DOF Hexapod (no servos/no electronics)	$248.90	LM PHOE
1	Arduino Mega 2560	$45.95	MS MKSP5; AF 191; SF DEV-11061
1	DFRobot mega sensor shield	$19.95	AZ B0098SJ1RS
6	HiTec HS-485HB servos	$16.99	LM S485HB; AZ B00944TF72

Count	Part	Estimated price	Part number(s)
6	HiTec HS-645MG servos	$31.49	LM S645MG; AZ B003T6RSVQ
6	HiTec HS-5685MH servos	$39.99	LM S5685MH; AZ B003X6KT7C
1	6V - 12V NiMH / NiCd smart charger	$21.95	LM USC-02; AZ B001DHC2LO
1	6V / 2800 mAH Ni-MH rechargeable battery	$26.95	LM BAT-05

Six Hitec HS-485HB servos are specified for the coxae (hip joints), six Hitec HS-645MG for the tibias, and six Hitec HS-5685MH for the femurs. Lynxmotion recommends 6 HS-485HB and 12 HS-645MG, but the HS-5685MH servos provide more torque for the femurs. You will want all of the torque that you can reasonably afford.

The list of required tools is modest. For the basic assembly, you will need:

- Small screwdriver set
- Needle-nose pliers
- Power drill
- 1/8" drill bit
- 1/4" drill bit
- Tie straps (a few dozen)

Controlling the Robot from the Command Line

Throughout construction, you will want to control the servos so you can align all of the joints with your coordinate system. Node's REPL is the perfect tool for this. Being able to send commands to a servo and see it respond in real time is pretty cool and useful.

> **REPL**
>
> REPL is short for read-eval-print loop. It gives you the ability to input things at the command line, then have code evaluate that input, and finally output the results and then wait for the next command.

To control this robot with the REPL, you will need the *phoenix.js* program:

1. Retrieve the *Buck.Animation/* directory from GitHub (*https://github.com/rwaldron/javascript-robotics*).
2. Drop the contents of this folder into a working directory on your computer.
3. Change directory to that folder, then use npm to install the dependencies in that directory:

```
npm install
```

This should give you everything you need for configuring and controlling your robot. If you do not already have Node.js installed on your computer, see "Installing Node.js".

An Introduction to phoenix.js

The JavaScript file, *phoenix.js*, holds all of the configuration and control code. Later in this chapter, you will need to set some configuration values in this code, but for now let's just read through the code and get familiar with some of the major features.

There are two major sections of the file where you may need to make edits. Near the beginning of the file, you'll find the first: the `Configurables` section. These objects describe the positions for all of the walking and turning steps of our hexapod. You should only have to edit these if you are using different servos than I have listed in the instructions.

The other section where you will definitely need to make edits is where we create the servo objects for our legs. This happens just inside the `board.on("ready")` callback. You will need to center your servos using the offset property for all 18 servos. I'll provide detailed instructions for this later. Also, if you are using different hardware, you may need to adjust the range properties of the servos.

A servo in the `phoenix` object is created like this:

```
phoenix.r1c = new five.Servo({
  pin:27, ❶
  invert: true, ❷
  offset: 0, ❸
  startAt: 90, ❹
  range: [50, 180] ❺
});
// + 17 more of these ...
```

❶ `pin` is the pin the servo is connected to on the DFRobot shield.

❷ `invert` will invert angles sent to the servo (i.e. 180 becomes 0, 45 becomes 135, etc.).

❸ `offset` trims the servo position.

❹ `startAt` is the starting position for the servo.

❺ `range` sets the max and min values that each servo will reach.

Table 5-2 shows which type of servo and which pin to use for each joint.

Table 5-2 *Servo table*

Servo	Abbreviation	Model	Pin
Left 1 Coxa	L1C	HS-485HB	27
Left 1 Femur	L1F	HS-645MG	26
Left 1 Tibia	L1T	HS-5685MH	25
Left 2 Coxa	L2C	HS-485HB	23
Left 2 Femur	L2F	HS-645MG	21
Left 2 Tibia	L2T	HS-5685MH	20
Left 3 Coxa	L3C	HS-485HB	19
Left 3 Femur	L3F	HS-645MG	18
Left 3 Tibia	L3T	HS-5685MH	17
Right 1 Coxa	R1C	HS-485HB	40
Right 1 Femur	R1F	HS-645MG	39
Right 1 Tibia	R1T	HS-5685MH	38
Right 2 Coxa	R2C	HS-485HB	49
Right 2 Femur	R2F	HS-645MG	48
Right 2 Tibia	R2T	HS-5685MH	47
Right 3 Coxa	R3C	HS-485HB	45
Right 3 Femur	R3F	HS-645MG	44
Right 3 Tibia	R3T	HS-5685MH	43

If you look closely at the code, you'll notice that the coxae on the rear legs flip their `invert` value relative to the other coxae servos on the same side. This is because you will want increasing values to move the coxa forward on the rear legs, but backward on all the others. Further down in the code, *phoenix.js* defines a `Servo.Array` for each leg:

```
phoenix.l1 = new five.Servo.Array([
  phoenix.l1c,
  phoenix.l1f,
  phoenix.l1t
]);
// Five more of these
```

When you call a method on a `Servo.Array` like `phoenix.l1`, that method is called with the same parameters on every Servo in the `Servo.Array`. We also use `Servo.Arrays` to group servos into animation targets. More on that later. *phoenix.js* also creates `Servo.Arrays` for:

coxa
: All six coxa servos

femur
: All six femur servos

tibia
: All six tibia servos

legs
: All 18 servos

Finally, *phoenix.js* creates an array of arrays for the joints. Those will be used later in the sleep and stand animations:

```
phoenix.joints = new five.Servo.Array([
  phoenix.coxa,
  phoenix.femur,
  phoenix.tibia
]);
```

You'll see that the *phoenix.js* file has created `Servo.Arrays` for many different combinations of servos. This upfront work makes creating animation segments easier. *phoenix.js* only has to send commands for those things that are unique.

Assembling the Robot

Before you can get started with the Animation class, you will need to assemble the robot. For most of the process you can follow the instructions provided by Lynxmotion, but there a few places where you will want to take extra care. The major steps for assembly are:

1. Prepare the chassis.
2. Mount the electronics.
3. Prepare the servos.
4. Install the coxae.
5. Install the femurs.
6. Install the tibias.

Prepare the Chassis

One problem that has to be solved is how to mount the electronics. Lynxmotion designed the Phoenix for a BotBoarduino. That board has mounting holes in different places than the Arduino Mega. Also, the Arduino Mega is big. Mounting it inside the Phoenix chassis is ideal but difficult. You will have to move the chassis offsets out enough to accommodate the Mega's width. For now, you can just mount the Arduino on top of the chassis.

Take the top chassis plate and place the Arduino on top. In Figure 5-2, you can see where we have used two of the small plastic offsets. The Arduino Mega's mounting holes will align with these offsets. Note that it's not a perfect fit. The Arduino will be at a slight angle relative to the chassis. If this bothers you, you can drill your own mounting holes.

Figure 5-2 *Assembled chassis*

From this point forward, the text will refer to the legs of the hexapod as R1 through R3 and L1 through L3.

Mount the Electronics

To mount the electronics, follow these steps:

1. Attach the Arduino to the 1/2" spacers.
2. Attach the shield to the Arduino.

3. The Phoenix comes bundled with a switch for the servo power. To mount the switch, you will need a 1/4″ hole. The holes that are pre-drilled in the chassis will be obscured by your electronics. Drill a new hole for your switch. I recommend installing it next to R2. Mount the switch onto the chassis as shown, trim the wires to length, and attach them to the power inputs on the shield.

4. The rechargeable Ni-MH battery fits snuggly between the chassis offsets. You won't need any mounting hardware. Insert the battery at a 45° angle and rotate it into place. Figure 5-3 shows the assembly so far.

Figure 5-3 *Chassis with electronics and power*

Prepare the Servos

Next, to prepare the servers, do the following:

1. Plug all of the coxa and femur servos into the Mega Sensor shield using Table 5-2 for reference. You do not need to align the tibia servos just yet, but if you want to plug them in and make sure they work, go ahead.

2. Plug your Arduino into your computer's USB port and attach power to your shield. Figure 5-4 shows how everything should look at this point.

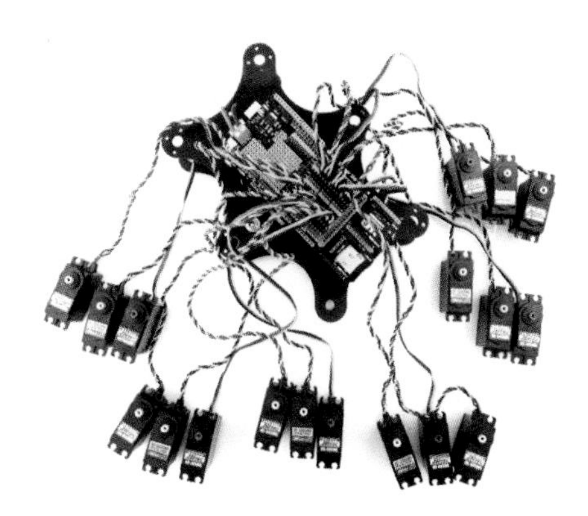

Figure 5-4 *All the servos connected*

3. You will want to align the servo horns as close to center as possible. To do this, you need to center the servos before pressing on the servo horns. In the directory where you put *phoenix.js*, run the following command:

```
node phoenix
```

4. Upon initialization, the servos will be set to their `startAt` values. You need the servos to all be at 90° so run the following command in the REPL:

```
ph.joints.to(90);
```

5. This will set all of the servos to 90°. You can now press a servo horn on with the mounting holes parallel to the axes of the servo. Only do this for the coxae and femurs. Do not press servo horns onto the tibia servos. When pressing on a servo horn, the holes should be aligned as shown in Figure 5-5.

6. Repeat this for all 12 coxa and femur servos. Tighten the servo horns only after you have disconnected the shield from its power source.

The coxae servos are threaded differently than the femur and tibia servos. Don't mix up the screws or you could ruin your servos.

Figure 5-5 *Correctly aligned servo horn with servos at 90°*

Installing the coxae

Follow these steps to install the coxae:

1. Assemble all six coxa/femur joint assemblies per the manufacturer's instructions. Note that three of the assemblies are mirror images of the others, as shown in Figure 5-6.

Figure 5-6 *Assembled coxae*

2. Now insert the two middle leg coxa/femur assemblies into the chassis (that's R2 and L2). Don't fasten the servo horns to the chassis yet.
3. Thread the servo wires between the DFRobot shield and the chassis.
4. Plug the four servos into the DFRobot shield. Use Table 5-2 to see where each one plugs in.
5. With the Arduino connected to your computer and the battery connected to your shield, run *phoenix.js* and set all of the joints to 90°. Note that you are only concerned with the coxae right now.
6. Rotate the coxae so that the femur servo horns are facing forward and the servo horn mounting holes line up with the holes in the chassis. The coxae might not be at a perfect right angle to the chassis yet. Don't worry—you will adjust that later.
7. Screw two of the servo horn screws through the chassis and into each servo horn, but do not tighten yet.
8. Disconnect the USB and the power, then tighten all four screws in each servo horn.
9. Repeat steps 1–8 for the front legs (R1 and L1). Notice that the mounting holes for the coxae servo horns are rotated 45°. The femur servos will be sticking out at the same 45° angle toward the front of your hexapod.
10. Finally, repeat steps 1–8 for the rear legs (R3 and L3). These mounting holes are also rotated 45°. The femur servos will be sticking out at a 45° angle toward the rear of the hexapod.
11. Run the *phoenix.js* code and attach the servo horns to the femur segments with the servos centered at 90°.
12. Disconnect the USB and power.
13. Fully tighten all four screws in each servo horn.

Figure 5-7 shows your work so far.

Figure 5-7 *Chassis with all six coxae*

Installing the Femurs

Follow these steps to install the femurs:

1. Run *phoenix.js* and set all of the servos to 90°.
2. The femur segments should already have servo horns mounted on them. Take the six femur segments and press them onto the femur servos. Do this for all six legs. The femurs will be sticking straight out, roughly parallel to the ground.
3. Stop *phoenix.js* with Control-C and disconnect the Arduino from your computer. It is now safe to screw the femur onto the femur servo. Figure 5-8 shows the installed femurs.

Figure 5-8 *All six femurs installed*

Installing the Tibias

You're almost done with the assembly, so all that's left are the tibias:

1. Mount the six tibia segments onto the tibia servos. Keep in mind that three are mirror images of the others.
2. Run the wires for the tibia and plug them into the shield, but do not attach the tibia servos to the femurs just yet.

When you run phoenix.js again, all of the joints on your robot are going to move. If it is sitting flat on the table it could flop around and make a serious mess of things. From this point forward, your hexapod needs to be off the ground when you run phoenix.js. You can hold it, or better yet build a test stand and use that. If you choose to hold your robot, be careful, it can pinch.

3. Run the code and set all of the servos to 90°.
4. Attach the tibia servos to the femurs. The end effectors should be beneath the tibia servo horn. Angle the tibia a few degrees inward, toward the chassis.
5. Disconnect the power and USB and screw the tibia servos to the femur.

Now is a good time to use tie straps to clean up your wires and fasten them in place. Remember to keep enough slack in the wires so that each of the limbs has a full range of motion. Do not leave so much slack that the wires can get caught up in the legs. Figure 5-9 shows the installed tibias.

Figure 5-9 *Tibias installed and all joints set to 90°*

The Coordinate System

For this project, you are going to use a separate, local coordinate system for each leg. Instead of defining the end effector positions with a three-tuple coordinate in X, Y, Z space, you will be using servo angles. This is known as a *joint coordinate system*. This is not as powerful as using inverse kinematics in a global coordinate system, but is better suited for learning the Animation class.

Because *phoenix.js* is using the joint coordinate system to create scripted movements, it explicitly sets the angle of each servo. It also orients the local coordinate systems so that the axes are parallel. Finally, *phoenix.js* inverts the direction of rotation between the right and left sides, and also between the front and back of the robot.

This arrangement makes animating the robot as easy as possible. If you plan on scripting your own custom animations, it is helpful to draw a coordinate system onto the base of your test stand. Set the origin point to the center of the chassis. Consider using a grid size of one centimeter on a test stand.

Trim the Servos

So far we've done our best to align the servo horns, but the splines limit the position of the output gear and each servo horn is likely to be a few degrees off. You can use the servo option's `offset` property to compensate for this.

The time you spend here calibrating all of the servos will simplify the process of scripting new animations. If the angle for stepping forward for all the coxae is exactly the same, you only have to calculate the movement one time. This upfront work will pay off in the long run.

Trim the coxae

First, let's adjust the l2c coxa:

1. Looking down at the robot from above (Figure 5-10), the coxa and femur should be at a right angle to the robot (parallel to the x axis on the coordinate grid). If you are using the test stand with a coordinate grid drawn upon it, this should be pretty easy to see. If the L2C servo needs to rotate clockwise, you will increase the angle from 90. To rotate counterclockwise, decrease the angle. Try incrementing a few degrees at a time and narrow your value down to the best fit. Don't bother with fractions of a degree, Johnny-Five only sends integer values to the servos:

```
ph.joints.to(91); // Get us closer
ph.joints.to(94); // Oops, too far
ph.joints.to(93); // Perfect!
```

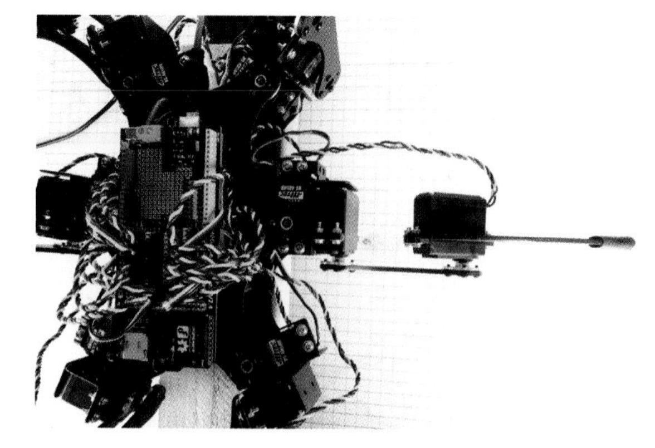

Figure 5-10 *Trimmed coxa*

2. Once you find the ideal value, you will need to subtract 90 from that number, which will give you your offset value. In this example, 93 – 90 = 3, so the `offset` is 3. Offset works exactly like the trim adjustment on an R/C radio. It will center servos that are off by a few degrees.

3. Find the line in *phoenix.js* where the `l2c` servo is created and adjust the `offset` value:

```
l2c: new five.Servo({
  pin:27,
  invert: true,
  offset: 3,
  startAt: 90,
  range: [50, 130]
})
```

4. Repeat steps 1 and 2 for the `r2c` servo, but remember, this servo is the inverse of the left side. You will need to invert the rotation angles.

5. Now examine the `l1c` servo. This servo is at an approximately 45° angle from the x axis. You need it to be parallel, so set the servo to 45°. That should get you pretty close, but you will need to fine-tune the position. The process looks something like this:

```
ph.joints.to(45); // Much closer
ph.joints.to(47); // Doh, wrong way!
ph.joints.to(44); // Perfect!
```

6. Once you are done, subtract 90 from the final angle. This will be your `offset` (a negative number).

7. Update the offset value for `l1c` in *phoenix.js*. It's not unusual for the offset value to be as high as 20°.

8. Repeat steps 4–6 for the other three coxae. Next, stop and restart *phoenix.js*. Now when all of the coxae are set to 90° they should be parallel to the x axis.

Trim the Femurs

The process of trimming the femurs is similar to the one we followed for the coxae. Again, you want the femurs to be parallel to the x axis when set to 90°:

1. Looking from the front or rear of the robot, adjust the femur servo offsets so the femurs are parallel to the ground.

2. Take that value, subtract 90, and that's your `offset`.

3. You will be setting the `offset` value for all six femurs: `l1f`, `r1f`, `l2f`, `r2f`, `l3f`, and `r3f`. Figure 5-11 shows a trimmed femur.

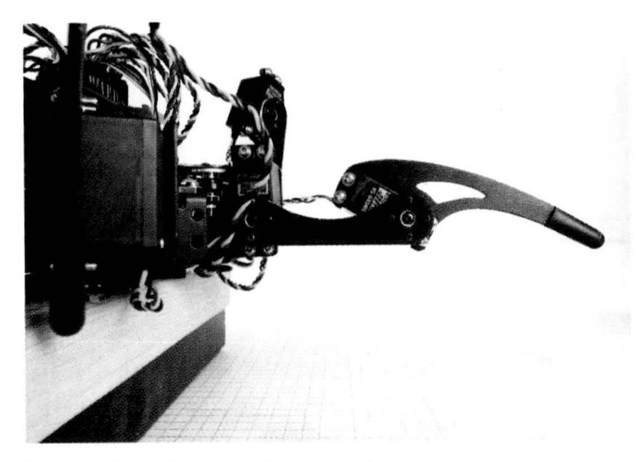

Figure 5-11 *Trimmed femur*

Trim the Tibias

Finally, you need to adjust the tibia. This time you want a line drawn from the center of tibia servo horn to the end effector to be vertical when the servo is set to 90°, as shown in Figure 5-12.

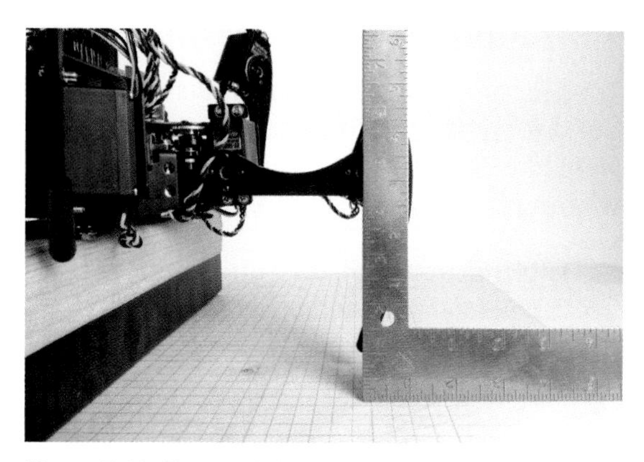

Figure 5-12 *Trimmed tibia*

When you find the offset values for the tibias, add them into the *phoenix.js* file for `l1t`, `r1t`, `l2t`, `r2t`, `l3t`, and `r3t`.

Add Ranges to the Servos

Now you will want to set the range of each servo. Many 180° servos do not have a full 180° range of motion. This is normal. You may be able to reprogram servos to recover this range, or you can insert a servo stretcher between the servo and the shield. You won't need the full range of each servo on your hexapod, so try not to worry about it. You actually want to further limit the range of your servos so that you don't overextend them to the point that the legs interfere with each other.

For this part, you want to control just one servo at a time:

1. Again, let's start with the L2 coxa. You want to rotate the coxa in each direction and find a practical limit. The servo will be able to rotate further than you would ever want it to.
2. Establish a limit in both directions through trial and error. If you reach a value where the servo no longer responds, you've exceed the servo's range of motion. For example, if the servo moves to 165, but does nothing when you move it to 166, your upper limit is 165. That process looks like:

```
ph.l2c.to(60);  // Still room
ph.l2c.to(58);  // That's a good
                // spot
ph.l2c.to(122); // I can see the
                // tibias will hit
                // each other
ph.l2c.to(119); // PERFECT!!
```

3. Find the line where the l2c servo is created in *phoenix.js* and update the range property using the values you just found:

```
l2c: new five.Servo({
  pin:27,
  invert: true,
  offset: 3,
  startAt: 90,
  range: [50, 130]
})
```

4. Repeat steps 1–3 for the other five coxae servos and update those range values: `l1c`, `l3c`, `r1c`, `r2c`, `r3c`.

   You are going to use a much wider range for the femurs and tibias, as they are less likely to interfere with each other. In fact, you will probably want to allow the entire available range for both the femurs and the tibias.

5. Experiment to find the point at which the servos no longer move. You can use the same warmer/colder approach you used to find the coxa limits. Keep in mind that all the femur servos should have the same (or similar) range.
6. Update the range values for all the femur servos: `l1f`, `l2f`, `l3f`, `r1f`, `r2f`, `r3f`.
7. All the tibia servos should also share the same range values. Find those ranges.
8. Update the range values in *phoenix.js* for `l1t`, `l2t`, `l3t`, `r1t`, `r2t`, `r3t`.

The benefits of finding the actual servo ranges may not be obvious. If you set a movement to extend to 175°, but the servo only goes to 165°, then the end effector is not where *phoenix.js* expects it to be, and the Animation thinks it is still running even when the servo no longer moves. That would not look right.

The l2 femur and tibia look like this when you finish:

```
l2f: new five.Servo({
  pin:22,
  invert: true,
  offset: -2,
  startAt: 180,
  range: [25, 165]
}),
l2t: new five.Servo({
  pin:21,
  invert: true,
  offset: 4,
  startAt: 180,
  range: [21, 159]
}),
```

Once you've set the `offset` and `range` for all 18 servos, it's time for the fun stuff. Let's make this robot move!

Walking Is Hard!

It is certainly possible to animate a robot without the `Animation()` class, but it's a lot of work. You might think that to move the robot forward you just put the end effector in contact with the ground and then sweep the coxae backward. The problem with that is the end effectors will describe a circle centered on the coxa servo's center of rotation, as shown in Figure 5-13. The legs will each have their own circles and will be working against each other. This gives you a very sloppy walk sequence.

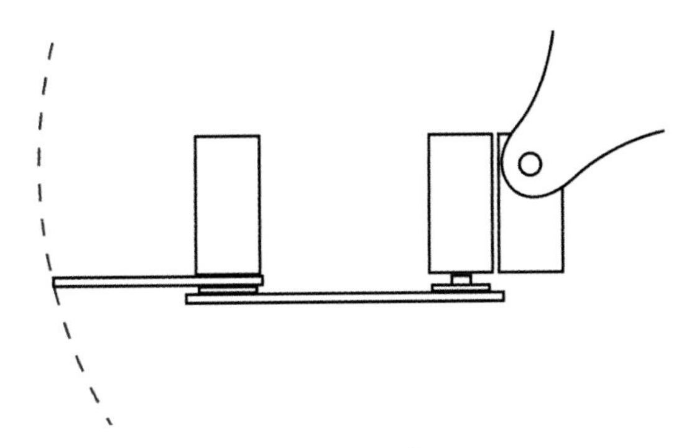

Figure 5-13 *This robot will look drunk*

To compensate for this, we need to move the tibia in and out to keep the end effector moving in a straight line, parallel to the direction the robot is moving, as shown in Figure 5-14. But wait! When you move the tibia in and out, the end effector moves up and down relative to the ground so we also need to move the femur up and down to compensate for that.

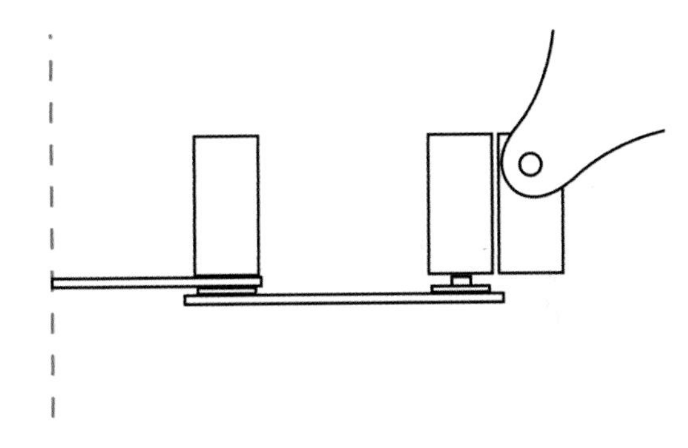

Figure 5-14 *This walking sequence will be smooth*

All 18 servos must be constantly working in concert to keep the robot's animation smooth. It can seem overwhelming, but Johnny-Five keeps it all manageable with the `Animation()` class.

Meet the Animation Class

`Animation.Class` is used to control robots with scripted movements. It serves as a wrapper for the Servo class. It handles timing, tweening, cue points, and key frames. The servo or group of servos you are animating is a target. The target can be an array of servos, a `Servo.Array`, or an array of servos and/or `Servo.Arrays`. That's confusing, so let's look at some examples.

An Array of Servos as the Target

In Example 5-1, the three servos of leg L1 are moved independently.

Example 5-1 *An array of servos*

```
board = new five.Board().on("ready",
                             function() {

  var l1c = new five.Servo(27);
  var l1f = new five.Servo(26);
  var l1t = new five.Servo(25);

  var myAnimation =
  new five.Animation([ l1c, l1f, l1t ]);

  // ...
});
```

This pattern allows you to pass a different value to each of those three servos at each `cuePoint` on the timeline.

A Servo.Array as the Target

Example 5-2 has the same result as the first example, but note that a single `Servo.Array` is passed as the default target. Again, the target is a single-dimensional array.

Example 5-2 *A Servo.Array*

```
board = new five.Board().on("ready",
                             function() {

  var l1c = new five.Servo(27);
  var l1f = new five.Servo(26);
  var l1t = new five.Servo(25);

  var l1 = new Servo.Array([ l1c, l1f,
l1t ]);

  var myAnimation = new five.Animation(l1);

  // ...

});
```

An Array of Servo.Arrays as the Target

In the final example (Example 5-3), the servos are grouped by joint type. This gives the ability to control the coxae as a group, the femurs as another group, and the tibias as a third group. Sending a value (angle) to a `Servo.Array` will send the same value to all the servos in the group.

Example 5-3 *An array of Servo.Arrays*

```
board = new five.Board().on("ready", function() {

  var l1c = new five.Servo(27);
  var l1f = new five.Servo(26);
  var l1t = new five.Servo(25);
  var l2c = new five.Servo(23);
  var l2f = new five.Servo(22);
  var l2t = new five.Servo(21);

  var coxa = new Servo.Array([ l1c, l2c ]);
  var femur = new Servo.Array([ l1f, l2f ]);
  var tibia = new Servo.Array([ l1t, l2t ]);

  // A two-dimensional array of arrays
  var myAnimation = new five.Animation([ coxa, femur, tibia ]);
```

```
  // ...
});
```

Being able to pass `Servo.Arrays` makes it possible to mix servos and `Servo.Arrays` in a single target. You can animate a `Servo.Array` just like any other device type.

Now look at the line in *phoenix.js*, where *phoenix.js* instantiates the animation and sets the default target:

```
var legsAnimation = new five.Animation(
phoenix.legs);
```

Here the default target, `phoenix.legs`, is a `Servo.Array` that consists of all 18 servos in the robot. This default target expects a value for every servo on the timeline at each `cuePoint`.

The First Animation Segment

Once you have an animation object, you can enqueue animation segments. An animation segment is a short modular sequence of movements. Segments are the place where all the angles, cue points, durations, and easing methods are defined. In short, they are where the magic happens.

Animation segments are synchronous. They run first-in/first-out through the animation's queue. When running *phoenix.js*, these animation segments can all be enqueued from the REPL. The following series of commands will make the robot stand, walk, stop, and sleep:

```
> node phoenix
ph.stand();
ph.walk();
ph.stop();
ph.sleep();
```

These are already defined for you in *phoenix.js*. You do not have to create or edit them. The following excerpts are shown here to help you understand how animation segments are formed.

Take a look at the first animation segment in *phoenix.js*, shown in Example 5-4.

Example 5-4 *Stand Animation Segment*

```
var stand = {
  target: phoenix.altJoints, ❶
  duration: 500, ❷
  cuePoints: [0, 0.1, 0.3, 0.7, 1.0], ❸
  oncomplete: function() { ❹
    phoenix.state = "stand";
  },
  keyFrames: [ ❺
    [null, { degrees: 90 }],
    [null, { degrees: 66 }],
    [null, false, false, { degrees: 120, easing: easeOut},
      { degrees: 94, easing: easeIn}],
    [null, false, { degrees: 106}, false, { degrees: 93 }]
  ]
};
```

❶ The `target` (optional) determines the servo(s) that are being animated. When the animation was created, `phoenix.legs` was passed in as the default target. For this segment, the default target is overridden with `phoenix.altJoints`. See Example 5-3 from earlier. By overriding the target, you can control different groups of servos. They are

Example 5-5 *phoenix.altJoints*

```
altJoints: new five.Servo.Array([
    phoenix.midCoxa,
    phoenix.outerCoxa,
    phoenix.femurs,
    phoenix.tibia
]),
```

all bound to the same timeline and animation queue. If you do not pass a target, the default target from the animation instantiation will be used for the segment.

❷ The `duration` (optional) specifies the duration of the animation in milliseconds. Note that adjusting the animation speed during playback will stretch or compress the duration. (Default: 1,000 ms.)

❸ `cuePoints` (optional) is a single-dimensional array of values from 0.0 to 1.0. Every animation segment has a timeline. On this timeline, you will define any number of `cuePoints`. Key frames are applied to devices at `cuePoints`. The `cuePoints/keyFrames` do not need to occur at regular intervals. With the duration of 500 ms in this example, the `cuePoints` are hit at 50 ms, 150 ms, 350 ms, and 500 ms. Changing the duration of an animation will scale the time that each `cuePoint` is hit (Default: [0, 1]).

❹ Animation will call this function when the segment has completed.

❺ The `keyFrames` are required, and are a two-dimensional array. The first dimension maps to the devices in the target and the length should be equal to the length of the target. To determine which device each element in the first dimension maps to, look at the code where `phoenix.altJoints` is defined, shown in Example 5-5 .

In `phoenix.altJoints`, there are four devices: `phoenix.midCoxa`, `phoenix.outerCoxa`, `phoenix.femur`, and `phoenix.tibia`. Each one of these is a `Servo.Array` of two, four, or six servos:

- `keyFrames[0]` maps to the inner coxae
- `keyFrame[1]` maps to the outer coxae
- `keyFrames[2]` maps to the femurs
- `keyFrames[3]` maps to the tibias

Each of the `keyFrame` arrays will have from 1 to *n* elements, where *n* is the number of `cuePoints`. Let's examine a couple of these arrays in the stand segment. First, let's look at `keyFrames[0]`, the innerCoxae:

```
[null, { degrees: 66 }],
```

`keyFrames[0][0]` defines the value for the first device in the target at 0ms. `keyFrames[0][1]` defines that value at 50ms. Any render performed between 0 ms and 50 ms will use a tweened value.

The first element in this `keyFrame` array is set to `null`. This indicates the animation should start from whatever the device's current position is. If the device is a `Servo.Array`, then the current position is read from the first member of that array. If you use `null` in any other position in the `keyFrame` array, the Animation will ignore the `keyframe` for that device. Only the first two `cuePoints` are defined in this array. All the remaining `cuePoints` will use the last known value (66).

Now look at `keyFrames[3]` (the tibias):

```
[null, false, { degrees: 106}, false, -13]
```

When `false` is used on a `keyFrame` element, Animation copies the previous element's calculated value. Setting the first two elements to `null` and `false` tells the animation to not move at all until the second `cuePoint` has passed.

When a number is passed as an element, Animation adds that number to the previous value.

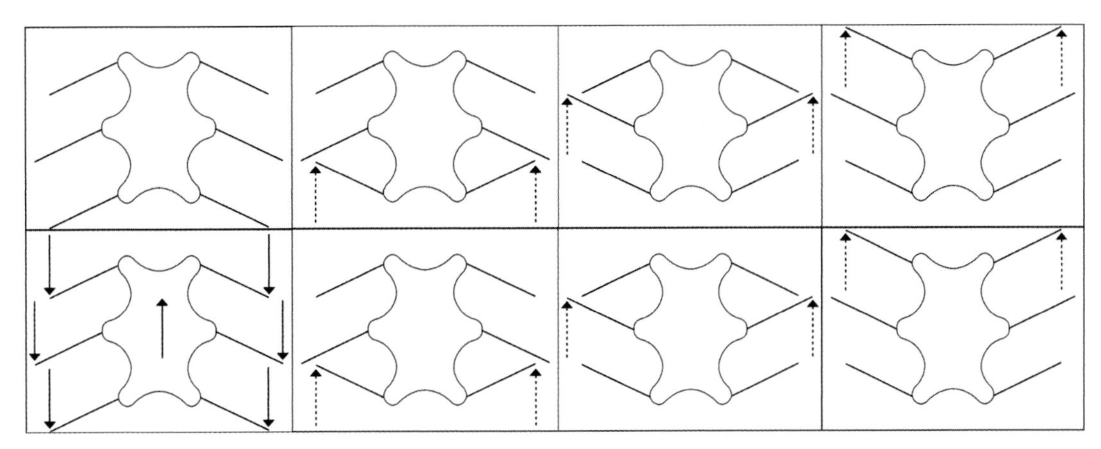

Figure 5-15 *The wave*

In this example, `keyFrame[3]` gets 106 from the prior frame and `keyFrame[4]` becomes 93.

This animation is designed to take the hexapod from its sleeping position to its home position (the home position being the assumed starting point for most animations).

If you compare the code in Example 5-5 to the code in phoenix.js you will notice that it is different. Instead of explicit number values in the code, phoenix.js is passing in elements of the `h` array. This array, defined at the top of the file, holds the position for the servos in each of the primary walking positions. Storing it in one place makes it easier to maintain and tweak later. Because using the array complicates understanding, numbers are substituted in the examples.

Walking

Robotic walking sequences are divided into two major categories: static and dynamic. Static gaits are any that keep the robot's center of gravity over a stable base (at least three legs) at all times. In other words, if at any time the robot were to stop moving mid-stride, it would not fall over. Dynamic gaits spend some time in a state of unbalance. *phoenix.js* only uses static gaits.

As stated earlier, hexapod walking sequences are more complicated than you might expect. You cannot just rotate the coxae to move a robot forward. Moving just the coxae causes the end effector to rotate about the coxa servo's center of rotation. The end effector needs to move in a straight line.

Moving the end effector in a straight line requires coordinated movements in all three servos of the leg. If the chassis is moving forward throughout the gait then all 18 servos must work together. They must maintain the position of the end effectors that are in contact with the ground. If you don't have all 18 servos working together, your robot will drag its feet and the gait will not be smooth.

There are four gaits defined in the *phoenix.js* file: row, crawl, walk, and run. For each of these gaits, an 8cm stride is used (the distance between the forward and rear end effector positions). Now let's look at each gait in more detail.

The Row Gait

The row sequence (more commonly known as wave) moves the legs forward in pairs and then moves the body forward. It's not what you'd call graceful, but it is an easy first gait and only requires eight keyFrame elements. See Figure 5-15.

It takes three `cuePoints` to define each of the leg movements. The first `cuePoint` is `null` or `false`, so the animation will start from the current position regardless of what that is. The second `cuePoint` defines the inflection point of the leg movement. The third `cuePoint` defines the end of the movement. Example 5-6 shows this.

Example 5-6 *The row gait*

```
var row = {
  target: phoenix.jointPairs,
  duration: 1500,
  cuePoints: [0, 0.1, 0.2, 0.3, 0.4, 0.5, 0.6, 0.7, 0.85, 1.0],
  loop: true,
  fps: 100,
  onstop: function() { phoenix.att(); },
  oncomplete: function() { },
  keyFrames: [

    [null, null, null, null, false, null, {degrees: 56},
      false, {degrees: 70}, {degrees: 91}],
    [null, null, null, null, false, { step: 30, easing: easeOut },
      {degrees: 116}, false, {degrees: 120}, {degrees: 119}],
    [null, null, null, null, false, { step: -20, easing: easeOut },
      {degrees: 97}, false, {degrees: 110}, {degrees: 116}],

    // ... two more leg pairs

  ]
}
```

The Walk Gait

The walk gait (Example 5-7) keeps four legs in contact with the ground at all times. The gait starts by moving one leg forward. When that leg reaches its inflection point, another leg is moved forward. The first leg should reach the ground at the same time the second leg reaches its inflection point. That's when the third leg begins to move.

A visual representation of the `keyFrames` for this movement would be hard to digest, so I've left that out. If you would like to examine how it works, change the duration of the segment to 20,000 ms and give it a run in slow-mo.

Example 5-7 *The walk gait*

```
var walk = {
  duration: 2000,
  cuePoints: [0, 0.071, 0.143, 0.214, 0.286, 0.357, 0.429, 0.5,
    0.571, 0.643, 0.714, 0.786, 0.857, 0.929, 1],
  loop: true,
  loopback: 0.5,
```

```
  fps: 100,
  onstop: function() { phoenix.att(); },
  oncomplete: function() { },
  keyFrames: [
    [null, null, {degrees: 82}, null, null, null, null, {degrees: 82}, null,
      {56}, null, null, null, null, {degrees: 82}], // r1c
    [null, { step: 30, easing: easeOut }, {degrees: 119, easing: easeIn}, null,
      null, null, null, {degrees: 119}, { step: 30, easing: easeOut },
      {degrees: s.f.f[0], easing: easeIn}, null, null, null, null,
      {degrees: 119}],
    [null, { step: -20, easing: easeOut }, {degrees: 119, easing: easeIn}, null,
      null, null, null, {degrees: 119}, { step: -20, easing: easeOut },
      {degrees: 97, easing: easeIn}, null, null, null, null, {degrees: 119}],

    ... five more legs
  ]
};
```

The Run Gait

The run gait (usually called tripod, shown in Example 5-8) gives the quickest movement, but it is hard on the femur and tibia servos. You should use the run gait sparingly. Run moves three legs at a time, leaving just three in contact with the ground. Those three legs carry more weight than legs in the other gaits.

Example 5-8 *The run gait*

```
var run = {
  duration: 1000,
  cuePoints: [0, 0.25, 0.5, 0.75, 1.0],
  loop: true,
  fps: 100,
  onstop: function() { phoenix.att(); },
  oncomplete: function() { },
  keyFrames: [
    [ null, {degrees: 70}, {degrees: 56}, null, {degrees: 91}],
    [ null, {degrees: 120}, {degrees: 116}, { step: 30, easing: easeOut },
      {degrees: 119, easing: easeIn}],
    [ null, {degrees: 110}, {degrees: 97}, { step: -20, easing: easeOut },
      {degrees: 116, easing: easeIn}],

  ... 5 more legs
  ]
};
```

Turning

Turning is harder than walking. Instead of a straight line, the end effectors will be describing concentric circles around the center of the robot. The radius of the circle for R2 and L2 will be one value and the other four legs will be another.

The turning gait is a lot like the run gait. We are keeping three legs in contact with the ground during each step and each step moves between 15° and 30°.

There are other animation segments included in *phoenix.js*, but not documented here: `sleep`, `waveRight`, `waveLeft`, and `crawl`. Once you have your robot calibrated and walking, give these other animations a shot.

Command Reference

Any of these commands may be called from the REPL while *phoenix.js* is running:

- `ph.stand()`
- `ph.sleep()`
- `ph.walk()`
- `ph.crawl()`
- `ph.row()`
- `ph.badRow()`
- `ph.run()`
- `ph.run("rev")`
- `ph.turn()`
- `ph.turn("left")`
- `ph.stop()`
- `ph.waveLeft()`
- `ph.waveRight()`
- `ph.att()`

Feel free to try them out and see what they do.

What's Next?

You should now have a hexapod that can walk, turn, and do a few other things on command, but this is only the beginning. You can use this platform to create new, even more awesome things. Here are just a few ideas for you to explore:

Create more animation segments
: Use a test stand and your imagination. Share them with other hexapod builders over at *http://forums.nodebots.io*.

Add a controller
: Having to type commands at the command line is a pain. You should explore options for controlling your hexapod. There are Node.js APIs for:

  - Leap Motion Controller (*http://bit.ly/19LY4nm*)
  - SparkFun Joystick shield—just use Johnny-Five
  - Kinect (*http://bit.ly/1bQRz4l*)
  - Wii Motion Controller—just use Johnny-Five
  - Brainwaves (OMG!) (*http://bit.ly/19LYoT4*)

Add sensors
: Use cameras and sonar devices to make your robot more aware of its environment. Teach it to make decisions based on that information.

Build a different hexapod
: The code here is not limited to the Phoenix hexapod. You can find many other similar robot chassis, and there are even open source designs for 3D printers from *http://www.thingiverse.com*. You'll need to update all the values in the `h`, `t`, `s`, and `l` arrays, but if you've built a test stand and aligned all your servos, this is relatively easy to do.

The decision to build this in Node.js and Johnny-Five really pays off. You have all of npm at your fingertips. This gives you a head start down whatever path you choose. What's more, Johnny-Five is the most flexible and best supported robotics API available, so you are free to explore any path you want to take.

Given the rise of the Maker Movement and the richness of the Node.js ecosystem, it's a pretty great time to be an amateur roboticist!

Building Voice-Controlled NodeBots

6

by Julián Duque

In this chapter, you will learn how to build a voice control for your NodeBot projects, from a basic controller using a microphone and loud sounds to a more advanced approach using speech recognition systems and an Android Wearable.

Initially, we will create a relay circuit that will be controlled with Johnny-Five, then we will create a microphone preamplifier circuit that will enable us to control the relay using loud sounds (see Figure 6-1). This approach will be useful if we are near our relay circuit. Then we will create a Node.js server that will respond to voice commands through real-time WebSockets and a REST API. At the end, we will see how to use other devices to interact with our NodeBot. For this project, we will be using a smart watch running Android Wear.

Our Johnny-Five code will control both circuits and will be integrated into our commands server. Both the electronics and the server will be managed by one single board. Other projects in this book use a computer to run Johnny-Five and the hardware board is somehow connected to the computer. Well, in this project, we will use a board that will allow us to run Node.js code in it: the BeagleBone Black.

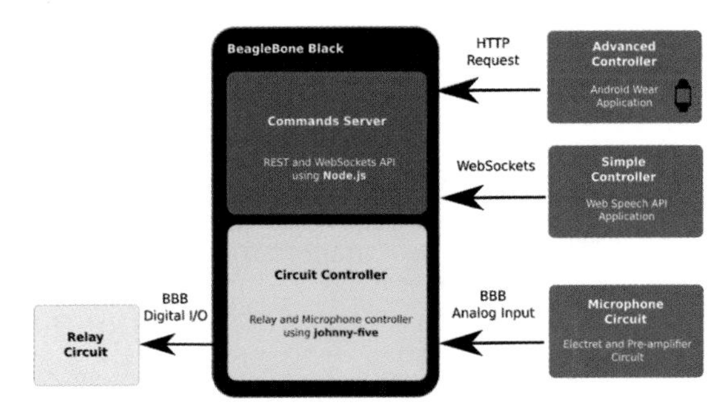

Figure 6-1 *Project architecture diagram*

Bill of Materials

We are going to use the Samsung Galaxy Gear Live, an Android smart watch for our advanced voice controller, but you can use any other Android Wear–capable device. Tables 6-1 through and 6-3 list the materials you will need.

Table 6-1 *Hardware*

Count	Part	Part numbers	Estimated price
1	BeagleBone Black	MS MKCCE4, AF 1996, SF DEV-12857	$54.95
1	Samsung Galaxy Gear Live	AZ B00LTR5HP6	$199.99

In this project, we are going to build a single relay circuit, but you can build as many as you need. The following parts are needed for one single channel. If you are planning to have more relays in your circuit, you will need to multiply the quantity by the number of relays you plan to use.

Table 6-2 *Relay circuit parts list*

Count	Part	Part numbers	Estimated price
1	5v / 110-220v Relay	SF COM-00100	$1.95
1	1N4001 diode	SF COM-08589	$0.15
1	2N3904 transistor	SF COM-00521	$0.50
1	1k resistor	DK 104669CT-ND	$0.10
1	Terminal block 3.mm (3 ports)	SF PRT-08235	$0.95
1	Solderable breadboard	SF PRT-12070, AF 571	$4.95
1	Jumper wires (male to male)	SF PRT-11026, AF 758, MS MKSEEED3	$4.95
1	Stick headers	SF PRT-00116	$1.50

The microphone circuit will be soldered along with the relay circuit, so if you are using more than one relay channel, you'll need a bigger protoboard.

Table 6-3 *Microphone circuit parts list*

Count	Part	Estimated price	Part numbers
1	Electret Microphone	$0.95	SF COM-08635
1	LM358 Op amp	$0.95	SF COM-09456
3	100nF capacitors	$0.75	SF COM-08375
3	1k resistors	$0.30	DK A104669CT-ND
2	10k resistors	$0.30	DK 104668CT-ND

Count	Part	Estimated price	Part numbers
3	100k resistors	$0.30	DK A105979CT-ND
1	4.7k resistor	$0.55	DK A105921CT-ND
1	220R resistor	$0.10	DK CF14JT220RCT-ND
1	10k potentiometer	$0.95	SF COM-09939
1	3mm Green LED	$0.35	SF COM-09650

You are going to need the following tools to solder the circuit:

- Soldering iron
- Alloy
- Helping hands
- Needle-nose pliers
- Wire cutter

If you connect your BeagleBone Black to your computer using a USB cable, it will serve as power supply and also will create a virtual network interface automatically. The IP address assigned to the BeagleBone Black is `192.168.7.2` *by default. If you are connecting your BeagleBone Black through Ethernet or WiFi, you'll need to define your own IP address and power it using a 5 V/1A DC power supply.*

BeagleBone Black

A BeagleBone Black (Figure 6-2) is a low-cost, community-supported electronics development platform that runs on Linux and has Node.js installed by default. It will take less than 5 minutes after the first boot to start developing Node.js applications. This board has both digital I/O and analog inputs, so it's perfect for our project. We will run our commands server along with the circuit controller in the BeagleBone Black.

Figure 6-2 *BeagleBone Black*

BeagleBone Black and Johnny-Five

BeagleBone Black default images (Ångström and Debian) runs a web-based IDE called Cloud9 (Figure 6-3). If you connected to your BeagleBone using a USB, you can navigate in your browser to *http://192.168.7.2:3000* and there you can start programming your board.

You can control a BeagleBone Black with Node.js using the official package `bonescript`. Take a look at the *demo/* folder in Cloud9 to see some examples.

In this project, we are going to use an I/O plugin for Johnny-Five called `beaglebone-io`, which is a wrapper on top of `bonescript`. This will enable us to use the same abstractions used in Johnny-Five without extra effort.

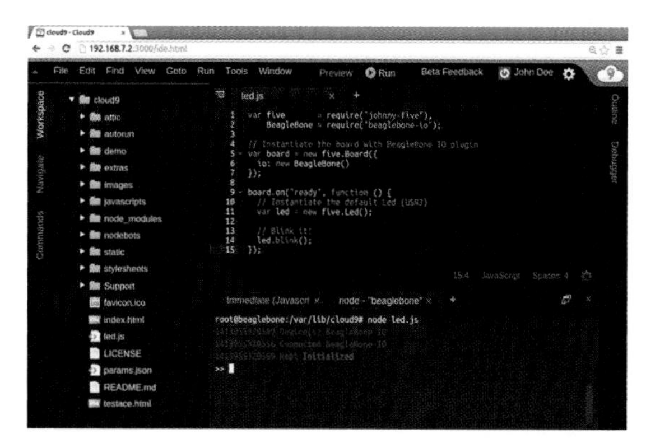

Figure 6-3 *Cloud9 running on the BeagleBone Black*

All source code for the examples in this book can be found at GitHub (*https://github.com/rwaldron/javascript-robotics*).

To install `beaglebone-io`, run the following command in the Cloud9 terminal:

```
npm install beaglebone-io
```

And then you can create a new file with the code shown in Example 6-1 to test if everything is working as expected.

Example 6-1 *Testing Johnny-Five with BeagleBone*

```
var five = require("johnny-five");
var BeagleBone = require("beaglebone-io");

// Instantiate the board with BeagleBone IO
plugin
var board = new five.Board({
  io: new BeagleBone()
});

board.on("ready", function() {
  // Instantiate the default LED (USR3)
  var led = new five.Led();

  // Blink it!
  led.blink();
});
```

The code will `blink` the `USR3` LED (Figure 6-4) in the BeagleBone Black.

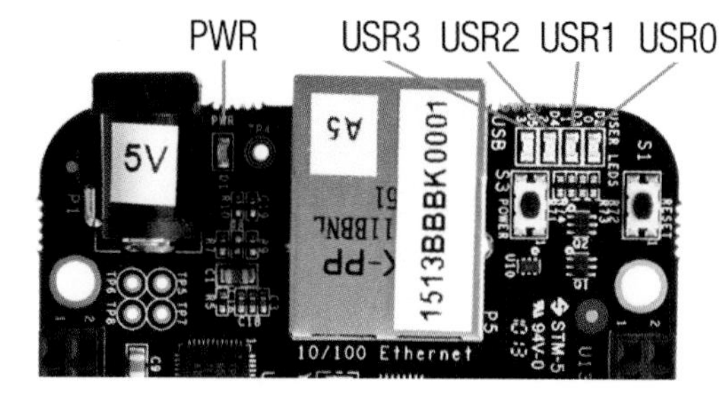

Figure 6-4 *BeagleBone Black LEDs*

Building Our Project

The following are the steps needed to build our voice-controlled NodeBots. In the following sections, we will cover:

- Building a relay circuit
- Building a microphone preamplifier circuit
- Building the commands server
- Simple voice control using the Web Speech API
- Advanced voice controller using Android Wearable

Building a Relay Circuit

Now that you have your BeagleBone Black working with Johnny-Five and `beaglebone-io`, it's time to build our Relay circuit. Given how BeagleBone Black only drives `3.3V` from the GPIO ports, you will need to use a transistor to increase the current to use a 5VDC relay.

You can build the circuit shown in the Fritzing schematic (*http://fritzing.org*) (Figure 6-5) on a protoboard but if you are planning to use it to control your house lights, it's better to use a solderable board or print it in a circuit. I'm going to show you what tools you need to build this circuit in a solderable board so it will be easier to use in your projects.

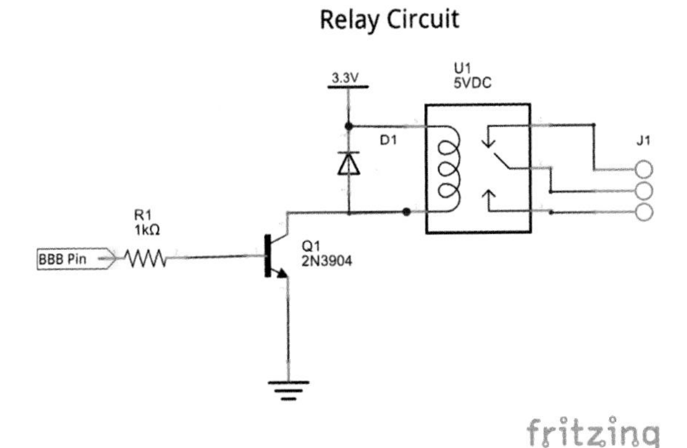

Figure 6-5 *Relay circuit schematic*

Figure 6-6 shows a picture of the finished circuit on a solderable breadboard.

Figure 6-6 *Relay circuit*

Controlling the Relay Circuit from Johnny-Five

This relay circuit will be controlled later using sounds and voice commands. We will want to turn on and off the relay switch from an external system. Because Johnny-Five is used to simplify the development of physical elements, there is an abstraction for a relay.

Our relay circuit will be connected to a digital port in the BeagleBone Black. We are going to wire it to the port `P8_8` and then we will write a basic Johnny-Five code to control the relay. The `relay` instance has two methods, `on` and `off`, which will be associated later with our commands server. Example 6-2 shows the source code for *relay.js*.

Example 6-2 *relay.js*

```
var five = require("johnny-five");
var BeagleBone = require("beaglebone-io");

var board = new five.Board({
  io: new BeagleBone()
});

board.on("ready", function() {

  // Pin 1 corresponds to P8_8 in the Beagle
Bone Black
  var relay = new five.Relay(1);

  // Turn on the Relay
  relay.on();

  // Turn off the Relay
  relay.off();

  // Inject relay instance to the REPL
  this.repl.inject({
    relay: relay
  });
});
```

To wire the relay circuit, connect the `3.3V` header pin on the circuit to port `P9_3` on the BeagleBone Black. Then connect the `GND` header pin to port `P9_1` so you can run it from the Cloud9 terminal by executing the following:

```
$ node relay.js
1415476684995 Device(s) BeagleBone-IO
1415476685028 Connected BeagleBone-IO
1415476685032 Repl Initialized
>>
```

At the Johnny-Five Repl prompt, you can execute `relay.on()` or `relay.off()` to test the relay circuit.

Now that the relay circuit is working, it's time to create our first controller using a microphone circuit.

Building a Microphone Preamplifier Circuit

For the first controller we'll play with, let's build a microphone preamplifier circuit (Figure 6-7).

Although this is not exactly voice control, it will be used to control the relay circuit using loud sounds like a clap, a whistle, or a shout. This circuit isn't needed for the voice control examples. This circuit is presented here as a basic control if you don't have access to Android Wear or if you want to build a simple clapper for your lights. All of the materials needed for this circuit are listed at the beginning of this chapter.

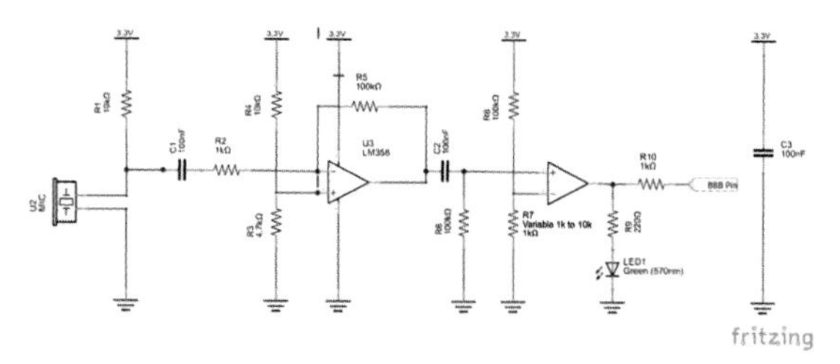

Figure 6-7 *Microphone preamplifier circuit schematic*

R7 must be connected in series with a 10k potentiometer. This is very useful to control the sensitivity of the microphone.

Figure 6-8 shows the finished circuit with the microphone preamplifier.

Figure 6-8 *Microphone circuit*

Connecting the Microphone to the Relay from Johnny-Five

Now that we have our two circuits assembled, it's time to put this together in our code. We will use BeagleBone Black pin `P8_8` (Digital Output) to connect the relay and `P9_39` (Analog Input) to connect the microphone circuit. This circuit will deliver a `Sound Pressure Level`. If this level reaches a maximum defined level, it will toggle the relay (from `on` to `off` and vice versa). The code will ignore changes in a timespan of 5 seconds (this time can be changed in code according to your needs).

Example 6-3 shows the circuit control code and sets up two elements, a `relay` and a `mic`, using the Johnny-Five abstractions `relay` and `sensor`. Input from `mic` is scaled to a range of `0` to `100`, and then we will compare the value with a defined `MAX_LEVEL`. If the value is higher than the `MAX_LEVEL`, we will toggle the value of the `relay`. We are using a flag called `detected` to avoid triggering the toggle. If multiple events are found in less than 5 seconds, this is defined in the `WAIT_TIME` variable.

Now you can try this basic control. Connect the microphone output to the analog input port `P9_39` on the BeagleBone Black. Try shouting, clapping, or making a loud noise near the microphone. You will see how the green LED will detect sound level and when the `MAX_LEVEL` is reached the relay will `click`. If this isn't working, try adjusting the `MAX_LEVEL` and `WAIT_TIME` values or change the microphone using the potentiometer.

Next, you will learn how to create a commands server using Node.js. This will be used along with the voice control that will be running on the BeagleBone Black.

Example 6-3 *Circuit controller (circuit.js)*

```
var five = require("johnny-five");
var BeagleBone = require("beaglebone-io");
var detected = false; // Flag to control sound level detection
var WAIT_TIME = 5; // Number of seconds before accepting a new change
var MAX_LEVEL = 95;

var board = new five.Board({
  io: new BeagleBone()
});

board.on("ready", function() {

  // Pin 1 corresponds to P8_8 in the BeagleBone Black
  var relay = new five.Relay(1);

  // Pin A0 corresponds to P9_39 in the BeagleBone Black
  var mic = new five.Sensor("A0");

  // Read sound level from mic, scaled from 0 to 100
  mic.scale(0, 100).on("data", function() {
    var level = this.value;

    if (!detected && level > MAX_LEVEL) {
      detected = true;

      // Toggle relay state
      relay.toggle();

      // Wait to prevent multiple toggles in a time frame
      setTimeout(function() {
        detected = false;
      }, WAIT_TIME * 1000);
    }
  });
});
```

Building the Commands Server

Our project is currently working with sound through a microphone circuit, but we want to add support to voice commands like `on`, `off`, `toggle`, or whatever we want to configure as a command. For this, we will need a commands server. We will create an `http` server with both a REST API and a WebSockets service to receive commands. This will enable real-time communications between the browser and the circuit controller. We will use `express` for our REST API and `Primus` for our WebSockets server.

Primus is an abstraction layer for real-time web applications; it's a wrapper for different WebSockets transports. You can check the documentation at the Primus GitHub (http://bit.ly/19LXkyw).

In your Cloud9 IDE terminal, run the following command to install the needed dependencies for our commands server:

```
npm install express primus primus-emitter ws --
save
```

The code shown in Example 6-4 can be placed in a file called *server.js* on your BeagleBone Black.

Example 6-4 *server.js*

```
var Primus = require("primus");
var PrimusEmitter = require("primus-emitter");
var express = require("express");
var http = require("http");
var path = require("path");

// Set up express static server and routes
var app = express();
app.use(express.static(path.join(__dirname, "public")));

//
// Receives a command via GET
// http://localhost:8080/command/{on,off}
//
app.get("/command/:command", function(req, res) {
  var command = req.params.command;

  // We have the command; we will need to send it to our circuit code
  // Implementation is shown later in this chapter

  res.send("ok");
});

// Create a basic http server for express
var server = http.createServer(app);

// Add WebSockets support to http server
var primus = new Primus(server);
primus.use("emitter", PrimusEmitter);

primus.on("connection", function(socket) {

  // If WebSockets server receives a `command` event, it will process it
  socket.on("command", function(command) {

    // We have the command; we will need to send it to our circuit code
    // Implementation is shown later in this chapter
  });
});

server.listen(8080);
```

Then create a folder called `public` in your project and put the code shown in Example 6-5 into a file called *index.html*. This file has a simple interaction with the WebSockets server using the `Primus` client. In the following example, it will send the command `off` after clicking the Send Command button.

Example 6-5 *index.html*

```
<!DOCTYPE html>
<html lang="en">
<head>
  <meta charset="UTF-8">
  <title>Voice Control</title>
  <script src="/primus/primus.js"></script>
</head>
<body>
  <button>Send command</button>
  <script>
    // Connect to the Web Socket server
    var socket = Primus.connect();

    var send = document.querySelector("button");

    send.onclick = function () {
      // Send command to the server using WebSockets
      socket.send("command", "on");
    };
  </script>
</body>
</html>
```

Run this server via Terminal by executing the following:

```
$ node server.js
```

You can also click the `Run` button in Cloud9 IDE. To test the example, you can go to the following URL: *http://192.168.7.2:8080*.

In this server and client, we are creating an `http` server using `express`. This server will respond to the following routes:

- `/commands/:command` where `:command` can be on or `off`
- A WebSockets server running using `Primus`
- *index.html* file contains a simple implementation of a WebSockets client

Next, you will learn how to integrate this client with a speech recognition system running in the browser.

Simple Voice Control Using the Web Speech API

We have a server ready to receive commands through WebSockets. Now we need to implement a way to talk to the server. For this simple controller, we will use the Web Speech API to achieve speech recognition from browser and mobile clients.

Web Speech API is an experimental specification and works only on WebKit-based browsers (Chrome, Safari, and Android). You can check Can I Use... (http://bit.ly/19LXnKV) to see if you can use the API in your favorite browser.

In our *index.html* file, we will add speech recognition support and we will send the final detected transcript back to the server through Web-

Sockets using the `command` event. Example 6-6 shows *index.html* with speech recognition.

Example 6-6 *Adding speech recognition*

```
<!DOCTYPE html>
<html lang="en">
<head>
  <meta charset="UTF-8">
  <title>Voice Control</title>
  <script src="/primus/primus.js"></script>
</head>
<body>
  <button>Send command</button>
  <h2></h2>
  <script>
    // Check if browser supports Web Speech API
    if ("webkitSpeechRecognition" in window) {
      var socket = new Primus.connect();
      var recognition = new webkitSpeechRecognition();

      // Get results inmediately
      recognition.interimResults = true;

      // Handle recognition result
      recognition.onresult = function(event) {
        var finalResult = "";

        for (var i = event.resultIndex; i < event.results.length; i++) {
          var result = event.results[i];

          // Get the final result
          if (result.isFinal) {
            finalResult = result[0].transcript;
          }
        }

        // Check if finaResult isn't empty
        if (finalResult) {
          document.querySelector("h2").innerHTML = finalResult;

          // Send result through WebSockets using the `command` event
          socket.send("command", finalResult);
          // Stop Speech Recognition
          recognition.abort();
        }
      };

      var send = document.querySelector('button');

      send.onclick = function() {
        // Start speech recognition
        recognition.start();
        document.querySelector('h2').innerHTML = "<em>Listening...</em>";
      };
```

```
    }
  </script>
</body>
</html>
```

The code starts the speech recognition system in the browser. After a command is recognized, it will be sent through WebSockets using the `Primus` client we created in the previous step.

Integrate the Commands Server with the Relay Circuit

Everything is ready to be put together. In this case, we will use an `EventEmitter` pattern to communicate both the commands server with the circuit controller. We will require a `circuit` object in our *server.js* file. This object will handle an event called `j5:command`, used to send a received command from the REST API or WebSockets to the circuit controller written using Johnny-Five. Example 6-7 shows the integrated *circuit.js*.

Example 6-7 instantiates a relay using Johnny-Five and exports an `emitter` object. This code will be imported in the *server.js* file, shown in the excerpt in Example 6-8.

Example 6-7 *circuit.js*

```
var five = require("johnny-five");
var BeagleBone = require("beaglebone-io");
var EventEmitter = require("events").EventEmitter;

// Create an emitter object to receive the commands from the server
var emitter = new EventEmitter();

var board = new five.Board({
  io: new BeagleBone()
});

board.on("ready", function () {

  // Pin 1 corresponds to P8_8 in the BeaglBone Black
  var relay = new five.Relay(1);

  // Receive the command from the server
  emitter.on("command", function (command) {

    // Check command received and execute an associated action
    if (command === "on") {
      relay.on();
      return;
    }

    if (command === "off") {
      relay.off();
      return;
    }

    if (command === "toggle") {
      relay.toggle();
      return;
```

```
    }
  });
});

module.exports = emitter;
```

Example 6-8 *server.js*

```
var Primus = require("primus"),
var PrimusEmitter = require("primus-emitter");
// Import the emitter from the Circuit module
var circuit = require("./circuit");

  // ...

var path = require("path");

// Create an event emitter object

// ...

app.get("/commands/:command", function (req, res) {
  var command = req.params.command;

  // Send the command to the circuit controller through
  // the event emitter
  circuit.emit("command", command);

  res.send("ok");
});

// ...

primus.on("connection", function (socket) {
  socket.on("command", function (command) {
    // Send the command to the circuit controller through
    // the event emitter
    circuit.emit("command", command);
  });
});
```

The preceding code uses the `circuit` emitter object and sends the command when received from the REST API or WebSockets event.

Advanced Voice Control Using an Android Wearable

With the appearance of wearables that can respond to voice commands, we can replace the Web Speech API part of the project with a device like Android Wear. This smart watch can be programmed to receive commands and interact with an existing API (in this case, with our commands server).

An Android Wear application consists of two applications: one installed on the mobile and the other installed on the wearable. The mobile application will be in charge of communicating with the commands server. All of the HTTP requests will be handled by this application. The wear application will be in charge of the speech recognition system and will send a `message` to the mobile using the Message API from Google Play services.

For more information about Android Wear, check out the Android Developer site (http://bit.ly/19LXmGC).

We will use Android Studio to develop these two applications. The installation instructions and how to create a project are found in the Appendix.

If you have the project created, then it's time to add some Java code to our project. If you don't know Java, don't worry—all of the code examples are ready to be added to the project.

Next, we will create an Android mobile application using the Android SDK.

Android Mobile Application

This application consists of a main activity (Example 6-9) with one text box and a button to test the communication between the mobile and the BeagleBone Black.

Example 6-9 *MainActivity.java*

```
package io.nodebots.voicecontroller;

import android.app.Activity;
import android.content.BroadcastReceiver;
import android.content.Context;
import android.content.Intent;
import android.content.IntentFilter;
import android.os.Bundle;
import android.view.View;
import android.widget.EditText;
import android.widget.TextView;

public class MainActivity extends Activity {

    // Class in charge of the HTTP Request
    private CommandRequest request;
    // Class in charge of the UI Update
    private BroadcastReceiver uiUpdated;

    @Override
    protected void onCreate(Bundle savedInstanceState) {
        super.onCreate(savedInstanceState);
        setContentView(R.layout.activity_main);

        // Instantiate the CommandRequest class using this Activity as context
        request = new CommandRequest(this);

        // This will be used to update the Last Command text from the
        // CommandRequest class
        uiUpdated = new BroadcastReceiver() {
            @Override
            public void onReceive(Context context, Intent intent) {
                TextView commandSent =
```

```
                (TextView) findViewById(R.id.commandSent);
            commandSent.setText(intent.getExtras().getString("command"));
        }
    };

    registerReceiver(uiUpdated, new IntentFilter("COMMAND_SENT"));
}

// Method associated to the Button, this is useful to send commands
// from the mobile without using the Wearable
public void sendCommand(View view) {
    EditText commandText = (EditText) findViewById(R.id.commandText);
    String command = commandText.getText().toString();
    request.doRequest(command);
    commandText.setText("");
}
}
```

All of the graphic interface is defined in the following two files: *activity_main.xml* and *strings.xml*. These files are created by default using Android Studio, so you'll need to update the content with the code shown in Example 6-10.

Example 6-10 *activity_main.xml*

```
<RelativeLayout xmlns:android="http://schemas.android.com/apk/res/android"
    xmlns:tools="http://schemas.android.com/tools"
    android:layout_width="match_parent"
    android:layout_height="match_parent"
    android:paddingLeft="@dimen/activity_horizontal_margin"
    android:paddingRight="@dimen/activity_horizontal_margin"
    android:paddingTop="@dimen/activity_vertical_margin"
    android:paddingBottom="@dimen/activity_vertical_margin"
      tools:context=".MainActivity">

    <EditText
        android:layout_width="wrap_content"
        android:layout_height="wrap_content"
        android:id="@+id/commandText"
        android:inputType="text"
        android:hint="Enter a command"
        android:layout_alignBottom="@+id/sendCommand"
        android:layout_alignParentLeft="true"
        android:layout_alignParentStart="true"
        android:layout_toLeftOf="@+id/sendCommand"
        android:layout_toStartOf="@+id/sendCommand" />

    <Button
        android:layout_width="wrap_content"
        android:layout_height="wrap_content"
        android:text="@string/send_command"
        android:id="@+id/sendCommand"
        android:clickable="true"
        android:layout_alignParentTop="true"
        android:layout_alignParentRight="true"
```

```
        android:layout_alignParentEnd="true"
        android:onClick="sendCommand" />

    <TextView
        android:layout_width="wrap_content"
        android:layout_height="wrap_content"
        android:textAppearance="?android:attr/textAppearanceMedium"
        android:text="@string/command_label"
        android:id="@+id/commandLabel"
        android:layout_below="@+id/commandText"
        android:layout_alignParentStart="true" />

    <TextView
        android:layout_width="wrap_content"
        android:layout_height="wrap_content"
        android:textAppearance="?android:attr/textAppearanceMedium"
        android:id="@+id/commandSent"
        android:layout_toEndOf="@+id/commandLabel"
        android:layout_alignTop="@+id/commandLabel"
        android:layout_alignEnd="@+id/sendCommand"
        android:textStyle="bold" />

</RelativeLayout>
```

Example 6-11 shows the static texts of the application.

The XML files in Examples 6-10 and 6-11 can be found in the *res/* folder in the Android mobile project.

Example 6-12 shows the *CommandRequest.java* file.

Example 6-11 *strings.xml*

```
<?xml version="1.0" encoding="utf-8"?>
<resources>

    <string name="app_name">Voice Controller</string>
    <string name="command">Command</string>
    <string name="command_label">Last command: </string>
    <string name="send_command">Send</string>
    <string name="action_settings">Settings</string>

</resources>
```

Example 6-12 *CommandRequest.java*

```
package io.nodebots.voicecontroller;

import android.content.Context;
import android.content.Intent;

import com.android.volley.Request;
```

```
import com.android.volley.RequestQueue;
import com.android.volley.Response;
import com.android.volley.VolleyError;
import com.android.volley.toolbox.StringRequest;
import com.android.volley.toolbox.Volley;

public class CommandRequest {

    private Context context;

    public CommandRequest(Context ctx) {
        context = ctx;
    }

    public void doRequest(final String command) {
        System.out.println("Executing request");
        RequestQueue queue = Volley.newRequestQueue(context);
        // BeagleBone Black IP Address
        // This will hit the /command/:comman route on the Node.js server
        String url = "http://192.168.1.10:8080/command/" + command;

        // Request a string response from the provided URL.
        StringRequest stringRequest =
          new StringRequest(Request.Method.GET, url,
                new Response.Listener<String>() {
                    @Override
                    public void onResponse(String response) {
                        // We will update the UI if the response is `ok`, this
                        // text is returned from the Node.js server
                        if (response.equals("ok")) {
                            // Update UI
                            Intent i = new Intent("COMMAND_SENT");
                            i.putExtra("command", command);
                            context.sendBroadcast(i);
                        }
                    }
                },
                new Response.ErrorListener() {
                    @Override
                    public void onErrorResponse(VolleyError volleyError) {
                        System.out.println(volleyError.networkResponse);
                    }
                }
        );

        // Add the request to the RequestQueue.
        queue.add(stringRequest);
    }
}
```

Example 6-12 will take care of the HTTP request between mobile and the commands server running in the BeagleBone Black. The IP address is hardcoded here, but you can create a textbox or settings options to change it from your mobile. We are using Volley, an HTTP client library used in Android Projects. The installation instructions are found in the Appendix.

Now the MainActivity and CommandRequest classes are ready. We will now create a ListenerService, shown in Example 6-13. This class will be in charge of receiving messages from the wear application.

Example 6-13 receives a message from the wear application and executes the doRequest method in the CommandRequest class.

Example 6-13 *ListenerService.java*

```
package io.nodebots.voicecontroller;

import com.google.android.gms.wearable.MessageEvent;
import com.google.android.gms.wearable.WearableListenerService;

public class ListenerService extends WearableListenerService {

    private static final String START_ACTIVITY = "/start/MainActivity";
    private CommandRequest request = new CommandRequest(this);

    @Override
    public void onMessageReceived(MessageEvent messageEvent) {
        if (messageEvent.getPath().equals(START_ACTIVITY)) {
            String command = new String(messageEvent.getData());
            request.doRequest(command);
        }
    }
}
```

Now the only thing needed to have this working is to add the respective permissions to our *AndroidManifest.xml* file. We need android.permission.INTERNET for the CommandRequest class and an intent-filter for our ListenerService. Example 6-14 shows this.

And it's done! You can test this application on your Android mobile device.

Android Wear Application

Next, you will implement the wear application, as shown in Example 6-15.

Example 6-14 *AndroidManifest.xml*

```
<?xml version="1.0" encoding="utf-8"?>
<manifest xmlns:android="http://schemas.android.com/apk/res/android"
    package="io.nodebots.voicecontroller" >

    <uses-permission android:name="android.permission.INTERNET" />
    <application
        android:allowBackup="true"
        android:icon="@drawable/ic_launcher"
        android:label="@string/app_name"
        android:theme="@style/AppTheme" >
        ...

        <meta-data android:name="com.google.android.gms.version" android:value="@integer/
google_play_services_version" />
```

```
        <service android:name=".ListenerService">
            <intent-filter>
                <action android:name="com.google.android.gms.wearable.BIND_LISTENER" />
            </intent-filter>
        </service>
    </application>

</manifest>
```

Example 6-15 *WearMainActivity.java*

```
package io.nodebots.voicecontroller;

import android.app.Activity;
import android.content.Intent;
import android.os.AsyncTask;
import android.os.Bundle;
import android.speech.RecognizerIntent;
import android.support.wearable.view.WatchViewStub;
import android.view.View;
import android.widget.TextView;

import com.google.android.gms.common.api.GoogleApiClient;
import com.google.android.gms.wearable.MessageApi.SendMessageResult;
import com.google.android.gms.wearable.Node;
import com.google.android.gms.wearable.NodeApi;
import com.google.android.gms.wearable.Wearable;

import java.util.Collection;
import java.util.HashSet;
import java.util.List;

public class WearMainActivity extends Activity {

    private TextView mTextView;
    private String node;
    private GoogleApiClient apiClient;
    private static final String START_ACTIVITY = "/start/MainActivity";
    private static final int SPEECH_REQUEST_CODE = 0;

    @Override
    protected void onCreate(Bundle savedInstanceState) {
        super.onCreate(savedInstanceState);
        setContentView(R.layout.activity_wear_main);
        final WatchViewStub stub =
          (WatchViewStub) findViewById(R.id.watch_view_stub);

        stub.setOnLayoutInflatedListener(
          new WatchViewStub.OnLayoutInflatedListener() {
            @Override
            public void onLayoutInflated(WatchViewStub stub) {
                mTextView = (TextView) stub.findViewById(R.id.text);
            }
        });
```

```
    apiClient =
     new GoogleApiClient.Builder(this).addApi(Wearable.API).build();
    apiClient.connect();
}

// Start Speech Recognizer
public void sendCommand(View view) {
    displaySpeechRecognizer();
}

// Get the connected nodes
private Collection<String> getNodes() {
    HashSet<String> results= new HashSet<String>();
    NodeApi.GetConnectedNodesResult nodes =
            Wearable.NodeApi.getConnectedNodes(apiClient).await();
    for (Node node : nodes.getNodes()) {
        results.add(node.getId());
    }
    return results;
}

// Create an intent that can start the Speech Recognizer activity
private void displaySpeechRecognizer() {
    Intent intent = new Intent(RecognizerIntent.ACTION_RECOGNIZE_SPEECH);
    intent.putExtra(RecognizerIntent.EXTRA_LANGUAGE_MODEL,
            RecognizerIntent.LANGUAGE_MODEL_FREE_FORM);

    // Start the activity, the intent will be populated
    // with the speech text
    startActivityForResult(intent, SPEECH_REQUEST_CODE);
}

// This callback is invoked when the Speech Recognizer returns.
// This is where you process the intent and extract the speech
// text from the intent.
@Override
protected void onActivityResult(int requestCode,
                                int resultCode,
                                Intent data) {
    if (requestCode == SPEECH_REQUEST_CODE && resultCode == RESULT_OK)
    {
        List<String> results =
          data.getStringArrayListExtra(RecognizerIntent.EXTRA_RESULTS);
        String spokenText = results.get(0);
        new SendMessageTask().execute(spokenText);
    }
    super.onActivityResult(requestCode, resultCode, data);
}

// Async Task used to send the message to the Mobile
private class SendMessageTask extends AsyncTask<String, Void, String> {

    @Override
    protected String doInBackground(String... command) {
        node = getNodes().iterator().next();
        SendMessageResult result =
          Wearable.MessageApi.sendMessage(
```

```
                apiClient,
                node,
                START_ACTIVITY,
                command[0].getBytes()).await();
            return result.toString();
        }

    }
}
```

Example 6-15 starts the speech recognition system after clicking the Send Command button. When the speech is recognized, it will send the command using an AsyncTask to the mobile through the Messages API from Google Play services.

More information about the Message API can be found on the Android Developer site (http://bit.ly/1O9dMcP).

Example 6-16 *rect_activity_wear_main.xml*

```
<?xml version="1.0" encoding="utf-8"?>
<LinearLayout xmlns:android="http://schemas.android.com/apk/res/android"
    xmlns:tools="http://schemas.android.com/tools" android:layout_width="match_parent"
    android:layout_height="match_parent" android:orientation="vertical"
    tools:context=".WearMainActivity" tools:deviceIds="wear_square">

    <TextView android:id="@+id/text" android:layout_width="wrap_content"
        android:layout_height="wrap_content" android:text="@string/welcome" />

    <Button
        android:layout_width="wrap_content"
        android:layout_height="wrap_content"
        android:id="@+id/commandButton"
        android:text="@string/command"
        android:clickable="true"
        android:onClick="sendCommand" />

</LinearLayout>
```

The interface is simple. We only have a text view and a button to start the speech recognition system. The file shown in Example 6-16 can be found in the *res/* folder in the Android Wear project.

The following line needs to be added to the *AndroidManifest.xml* of the wear project in order to use the Google Play service's APIs:

```
<meta-data android:name="com.google.an
droid.gms.version" android:value="@integer/
google_play_services_version" />
```

Now both projects are ready to be tested. Install both applications on your mobile and wear devices, and start controlling your home lights or appliances using voice commands!

What's Next?

Using the commands server we built, you can change the relay circuit with a robot rover or a drone and receive commands such as `forward`, `reverse`, `right`, `left`, and `stop`. You can also change the colors of an RGB LED. Be creative and build your next voice-controlled NodeBot!

7 An Indoor Sundial

By Lyza Gardner

It's entirely absurd to design and construct an indoor sundial like the one shown in Figure 7-1, which is why it's so fun to do so. We'll craft our sundial as a tongue-in-cheek nod to the astronomical timepieces that have existed for millennia.

Unlike traditional sundials, which require a clear day and an outdoor location, we'll use our own "sun" and position it as needed to tell, roughly, our own local solar time.

A horizontal-style sundial is typically disk shaped, constructed with a hand or pointer, called the *gnomon*, which is aligned to true north (or due south in the southern hemisphere). The gnomon rises away from the center of the disk at an angle equivalent to the local latitude. Thus, a sundial at higher latitudes has a steeper gnomon. The sun casts a predictable shadow over the gnomon, allowing us to tell local solar time based on the position of the shadow.

The Earth's orbit around the sun is wobbly and imperfect, and the clockwork we're constructing won't have the elegant perfection of a true timepiece. But it's still fun to have an approximation of the local time and of the sun's current position in our own sky. And it works even when it's cloudy!

Figure 7-1 *What time is it in Ireland? The sundial can tell us*

Bill of Materials

To build our sundial, we'll need some tools and materials. Building the sundial will involve some construction. We'll use foam core, which is a simple material to work with and provides the rigidity and structure we'll need. Table 7-1 lists the electronic parts and Table 7-2 lists the tools and supplies.

Table 7-1 *Electronic parts*

Count	Part	Notes	Estimated price	Part numbers
1	Arduino Uno	I've designed the sundial using an Arduino Uno, Rev. 3. You could certainly use a different Arduino if you wanted.	$25	MS MKSP99; AF 50; SF DEV-11021
1	Half-size, adhesive breadboard	Approx. 3.3 x 2.1″	$5	MS MKKN2; AF 64; SF PRT-09567
1	Standard servo	Standard-sized servo, 4.8-6.0v	$10	MS MKPX17; AF 155; SF ROB-09347
1	Sub-micro servo	Sub-micro servo, 4.8-6.0v	$9	SF ROB-09065
1	LED, white	Individual LEDs are cheap, but you may need to buy an assortment (e.g., MS MKEE7)—they're hard to find in single units!	$0.35	MS MKEE7
1	220-ohm resistor	1/8- or 1/4-watt is fine	A few pennies	any electronic parts supplier
1	Four AA battery holder	A wired holder for 4 AA batteries (6v) to drive the standard servo	$2	AF 830; PRT-12083

Table 7-2 *Tools and supplies*

Count	Part item	Notes	Estimated price
32 × 40″ sheet or equivalent	3/16″ (or 5mm) foam core	I used black, but white works, too. Most parts can be cut from letter- or A4-sized sheets, but the outer elevation arc is 9″ in diameter.	$5-6 for a 32 × 40″ sheet
10 sheets or so	Card stock	For aesthetic reasons, I used metallic-coated card stock in gold and silver.	White card stock: $.10/sheet Metallic card stock: $11/25 sheets

Count	Part item	Notes	Estimated price
1	Craft knife	X-ACTO or similar, with plenty of fresh blades	$4.50
1	Circle-cutting guide	You can use a purpose-built circle cutter, circle templates, or a compass and a steady hand. Circle diameters range from 6 to 9 inches.	$15-30 for a circle cutter; $12 for a compass
1	Protractor		$4
1	Ruler	Ideally cork-backed for guiding cuts	$4.29
	Adhesive	Hot glue gun (ideal), craft glue or a *strong* glue stick	$6.50-8.50 for a glue gun; $1 or less for a glue stick
1 pair	Wire cutters/ strippers		$11.50
	Stiff wire or metal rods	I used 1/16" brass rods, which I found at a local hobby-train shop. Memory wire (as used in jewelry making and readily available at craft stores) works OK, too.	$5
1	Small screwdriver		$4
1	Awl	Something sharp and pointy for marking and finding the center of circles	$4
1	Cutting mat	A self-healing cutting mat or similar will make your life easier	$15
Assorted	Spacers, screws	For mounting the Arduino (recommended)	
1 roll	Electrical tape		
1	Soldering iron	For wiring up the LED (recommended)	$14 to several hundred dollars

Foam Core Structural Pieces

Tables 7-3 through 7-5 list of all the pieces we'll cut out from foam core as we work through the process of building the sundial, listed in the order we'll cut them as we build.

Table 7-3 *Pieces we'll cut for the sundial base*

Count	Dimension	Name	Notes
1	8.5 × 5″	Sundial base	This is the base for the whole sundial. It will support other pieces, contain the AA battery holder, and allow for wire management.
1	4 × 7″	Mounting board	This will hold the Arduino, breadboard, and one of the servos.
4	1 × 3.5″	Mounting board supports	These will go on the sundial base and will support the mounting board.
3	2 × 1″	Smaller supports	
3	2 × .5″	Support reinforcements	

Table 7-4 *Pieces needed for the sundial disks*

Name	Diameter	Notes
Gnomon disk	6″	This 6"-diameter circle holds the gnomon. It is glued to the base disk and remains fixed, providing an inner guide for the azimuth ring to rotate around.
Azimuth ring	Inner 6 1/8″, outer 8″	This 8"-diameter ring will sit on the base disk and rotate around the gnomon disk.
Base disk	8″	This 8"-diameter circle will remain fixed and will support our azimuth ring and gnomon disk. A track arc inside of the disk allows our azimuth servo (with attached arm) to rotate the azimuth ring from below.
Elevation ring	inner: 8 1/16″, outer: 9″	Actually an arc, this 9"-diameter ring pivots on the azimuth ring to raise and lower the sun to the correct angle in the sky.

Table 7-5 *Pieces we'll cut for the base walls and supports*

Count	Dimension	Notes
1	7.5 x 4.5″	Rear wall
2	2.25 x 4.5″	Side walls
6	.25 x 2″	Support reinforcements
1	12 x 4.5″	Base disk support wall
1	8″ diameter semicircle	Base disk support guide
1	9.5 x 3.5″	Base disk support foot

Table 7-6 *Pieces for the azimuth arm*

Count	Dimension	Notes
1	3.75″ x 7/8″	Azimuth arm
3	Small/as needed	Square pieces to help support the azimuth arm wire

Building Our Sundial

Put on your crafting hat and Node.js shoes—it's time to build and code for your sundial. Perhaps, like me, you'll find that construction with foam core and a hot glue gun is peculiarly satisfying! To build the sundial, you'll need to:

1. Assemble a core mounting board and arrange components.
2. Wire and configure servos.
3. Assemble the lower base.
4. Make the disks.
5. Build the walls and disk supports.
6. Complete construction and hardware.
7. Write code to make it work.
8. Put it all together.

Cutting and Assembling the Core Structure

The base layout pieces can be cut from a single letter or an A4-size sheet of foam.

Assemble the Mounting Board

This piece will hold the Arduino, breadboard, and one of the servos that will drive our sundial. You'll use foam-core pieces from Table 7-3.

The standard-size servo mounted here will provide the rotation for your azimuth disk. It will place the LED "sun" in the correct angle around the circle of the horizon, where 0° is north, 90° is east, 180° is south, and 270° is due west. Here are the steps you'll need to follow:

1. Find the center of the (4″ × 6.5″) mounting board. Mark this spot.
2. Align your standard-size servo such that the shaft is directly above the center of the board and the long side of the servo is aligned with the short side of the mounting board. Trace lightly around the servo with a pencil to mark the location.
3. Take one of the 2″ × 1″ pieces cut earlier and place the servo on it. Trace carefully around the bottom of the servo, creating an outline on the foam core.
4. Cut this shape out from the center of the piece, creating a sort of *servo cozy* that should fit around your servo snugly. You may need to cut an escape notch for the servo's wires, depending on the exact shape of your particular servo.
5. Position the servo cozy on the board to match the traced outline and glue in place. The servo should fit snugly and

not wiggle easily. If you really want to make that servo secure, and you're ready to commit this servo to this project forever, you can hot glue it to the mounting board.

6. With the servo wires facing toward you, mount the Arduino on the right side of the mounting board, with the USB and barrel jack aligned along the short edge of the board. Mount the Arduino slightly rear of center, leaving about 1/3" clearance from the long, back edge of the board.
7. Mount the mini breadboard on the opposite side of the servo, approximately centered in the available space. See Figure 7-2.

Figure 7-2 *The mounting board showing the servo cozy and components*

Centering the servo shaft

Note that the shaft is not centered on servos, so your standard-sized servo will be off-center front to back on the mounting board when the servo's wires are facing you.

Wiring and Configuring the Servos

We've placed the azimuth servo on the mounting board, and later we'll use a second, sub-micro-sized servo to set the elevation of the sun (i.e., its height from the horizon in degrees).

Let's make sure our servos are ready to play their parts by wiring them as in Figure 7-3.

A standard servo sometimes needs more power than an Arduino's onboard power can reliably provide, so we'll want to hook its power and ground wires up to our external 6-volt AA battery holder. To power our sub-micro server, as well as our LED sun, we'll be using the Arduino's 5V power.

Wiring the Servos

Let's wire both of the servos.

Wiring the standard (azimuth) servo

To wire the standard (azimuth) servo, follow these steps:

1. Run 6V power and ground to one power rail of the breadboard.
2. Hook up the standard servo to the 6V power.
3. Connect the data wire to pin 9 on the Arduino.

Wiring the sub-micro (elevation) servo

To wire the sub-micro (elevation) servo, follow these steps:

1. Connect the Arduino's 5V power and ground to the other power rail.
2. Hook up the sub-micro servo to the 5V power.
3. Connect the data wire to pin 10 on the Arduino.
4. We need to have a common ground, so connect the grounds from each power rail to each other.

Configuring the Servos

We're going to turn to the coding side for a bit, so that we can get our servos set up. You can find the source for the sundial project in the *Gardner.Sundial/* directory on GitHub (*http://bit.ly/19LX9n3*). In the following sections, we'll install project dependencies and run the *servos.js* script to test our our servos.

All source code for the examples in this book can be found on this book's GitHub page (*http://bit.ly/19LX9n3*).

servos.js

The project script *servos.js* configures our servos. See Examples 7-1 and 7-2.

Example 7-1 *The servos.js script first creates johnny-five Servo objects for each of our servos*

```
var azimuthServo = new five.Servo({
  center: true, ❶
  isInverted: true, ❷
  pin: 9
});
var elevationServo = new five.Servo({
  center: true, ❶
  isInverted: true, ❷
  pin: 10
});
```

❶ Center each of the servos in its range (90 by default).

❷ Flip the rotation of the servo to be clockwise.

Example 7-2 *The servos.js script then makes a reference to those Servos available in the REPL*

```
this.repl.inject({
  aServo: azimuthServo, ❶
  eServo: elevationServo ❷
});
```

❶ aServo is the azimuth (standard) servo.

❷ eServo is the elevation (sub-micro) servo.

Running the code

When you're ready to run the code, follow these steps:

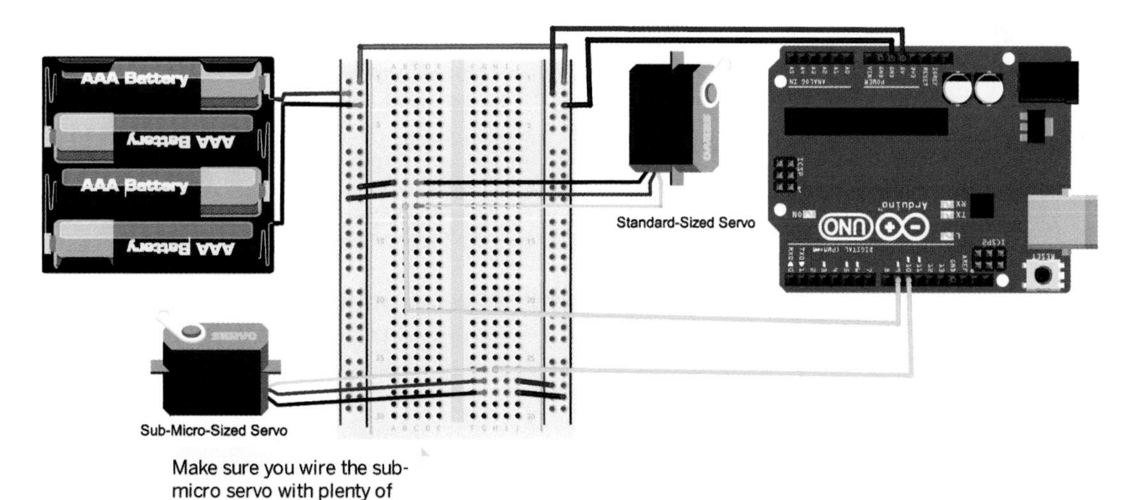

Figure 7-3 *Servo wiring Fritzing diagram*

1. Download or clone the *Gardner.Sundial/* project directory from GitHub (*http://bit.ly/19LX9n3*).
2. Install project dependencies. From within the project directory, run:

   ```
   npm install
   ```

3. Make sure your Arduino is connected to your computer's USB port, and execute the script by typing:

   ```
   node servos.js
   ```

After the board initializes, you should hear and see the two servos adjusting position to center (90°).

You can interact with the servos. Try typing things like:

```
>> aServo.to(120)
```

or:

```
>> eServo.to(50)
```

Although servos are typically described as having a range of 0°–180°, in actuality they usually have ranges of about 165°. That means you won't be able to get your servos to go all the way to 0° nor all the way to 180°.

Experiment in the REPL with each servo, moving it closer to 0 using the `to` method until it starts making funny noises or refuses to comply. Note the value you were able to get to (in my case, for the standard servo, it was 7). Do the same at the high end of the range, getting as close as possible to 180. Take note of these ranges for later.

The `to` method on Johnny-Five's `Servo` class does what it sounds like: moves the servo to that position. Learn more about `Servo` on the `johnny-five` (*http://bit.ly/1bQONfn*).

Building the Lower Base

Now that we have our servos ready to go, let's get back to the hardware side and construct a lower base so our sundial components have something to stand on. The base will also contain our azimuth servo's battery pack and provide some room for wire management. Fun!

1. Working with the 8.5″ × 5″ base piece from Table 7-3, pencil a line 3/4″ inset from the long edges and 15/16″ inset from the short edges so that you create an inner box shape representing the footprint of the inner base.
2. Glue two of the 3.5″ × 1″ support pieces on edge, centered within the *inside* of the 15/16″ lines on the short edges. This will make 1"-high supports at right angles to the base along its shorter edges.
3. Cut each of the smaller support reinforcements (2″ × .5″) in half lengthwise, which will result in six, 1/4"-wide reinforcement strips. Glue one reinforcement strip along the bottom *inside* corner of each support, strenghtening the structure.
4. Glue the two additional 3.5″ × 1″ supports at right angles to the first, as shown in Figure 7-4.
5. With one of the long edges of the sundial base facing you as in Figure 7-4, position the battery holder in the inside, rear right corner, with the battery holder's wires coming toward you. Mark the position of the battery holder's left edge and glue a shorter, 2″ × 1″ support there to create a little house for your batteries. It doesn't need to be too snug or accurate.

6. Flip the inner mounting board over, such that the Arduino is upside down and at the top left. Take two remaining 1/4″ × 2″ support reinforcements and glue them along the very edge of each of the short edges. Finally, take the remaining 2″ × 1″ support and glue it along the back, long edge of the flipped-over mounting board, toward the left side. These three little supports will help your mounting board stay in place on the lower base.
7. Flip your mounting board back over, and you'll be able to position it on the lower base with a satisfyingly snug fit if your cutting and gluing were accurate.

Figure 7-4 *The completed base construction, with the mounting board upside down*

Making the Disks

Our sundial needs its disks now! You'll find that cutting these is easier if you have a circle-cutting tool, or circle-shaped templates, but the disks in Table 7-4 can certainly be drawn with a compass and cut with a steady hand.

You'll need two pieces of foam core to cut these disks. The gnomon disk and azimuth ring can be cut from one piece, concentrically. Similarly, the circle base and elevation ring can be cut from a single piece:

1. Using a compass or circle template, draw a 6″ circle on foam core, making sure to note the center carefully with an awl, compass point, or something sharp and pointy. This will be the *gnomon disk*.
2. From the same centerpoint, draw a (concentric) 6 1/16″ circle, and then again from the same center, draw another, 7 1/2″ circle. This will be the *azimuth ring*.
3. Cut out the gnomon disk and azimuth ring.
4. In the gnomon disk, cut a slot from the center to one edge that is exactly the width of the foam core (3/16″). This slot will hold the gnomon later.
5. Draw an 8″ circle on a new sheet of foam core. Within this circle, from the same centerpoint, cut two 180° arcs on it that will create a track 1/4″ wide. The inner arc should be 7″ in diameter (3.5″ radius) and the outer 7.5″ in diameter (3.75″ radius). You can see this track in Figure 7-5.
6. Draw the elevation ring—inner diameter 8 1/16″, outer diameter 9″—from the same centerpoint as the base disk. Cut out the base disk and the elevation ring.

Optionally, cut out an additional 7″-diameter circle from decorative card stock (I used gold-foil card stock). Glue this to the center of the base disk, keeping its outer edge inside of the inset track, before gluing the gnomon disk to the base disk.

7. Glue the gnomon disk to the base disk, orienting the gnomon slot so that it

points *exactly* opposite from the base disk's track, as in Figure 7-5.

Figure 7-5 *The gnomon and base disks, showing the inner track and decorative extra circle on the base disk*

Build the Base Walls and Base Disk Support

We have disks, and now we need some walls and supports to put them on, using the pieces in Table 7-5. First, we'll build some walls, then we'll align the disks and build a support for the base disk.

Build the Base Walls

To build the base walls, follow the steps:

1. Glue a 1/4″ × 2″ support reinforcement along the long edge of the 7″ × 4.5″ rear wall, and along one short edge each of the 2.25″ × 4.5″ side wall pieces. See Figure 7-6.
2. Position and glue the rear wall flush along the rear long edge of the sundial base. There will be a gap between the rear wall and the mounting board. This is intentional for wire management as needed. The rear wall will extend beyond the length of the existing inner base by the width of the foam core (3/16″ on either end).
3. Position the side walls along the short edges of the inner base, abutting the new rear wall. Before gluing the side wall for the side that contains the overhanging Arduino jacks, cut a hole in the wall for the Arduino's barrel jack. You may also need to trim the side wall slightly narrower if the Arduino's USB jack gets in the way.
4. Glue 'em up! It should look like Figure 7-7.

Figure 7-6 *The three wall pieces with support reinforcements attached*

Figure 7-7 *The rear and side walls in place*

Position the Disks

It's important that we align the base and gnomon disks correctly in relation to the sundial base and walls:

1. Position the base disk atop the base walls. It will be front-heavy at this point.

2. Point the gnomon slot *exactly* perpendicular to the rear walls—the long walls of the sundial represent an east-west axis and the gnomon slot points "north."
3. Make sure the base disk inner track clears both side walls. The semi-circle track should be positioned to the front of the sundial.
4. Use a thin rod or awl through the base disk's center to align it with the shaft of the azimuth servo, as in Figure 7-8.

Figure 7-8 *Finding the center and placing the guides for the base disk*

5. With the disk positioned and centered, carefully mark with a pencil along the underside of the base disk along the back wall.
6. Glue one of the support reinforcements accuraely along the *outside* of the mark on the underside of the base disk. This support should serve as a guide along the outside edge of the back wall for placing the base disk.
7. Replace the base disk on the back wall, using the new guide. Find the circle center again, and make sure the circle is centered, left-to-right, over the servo shaft.
8. Mark the underside of the base disk along the *inside* of the two side walls. Glue two additional supports along the inside of those marks, as in Figure 7-9. These disk supports will be positioned on the *inside* of the short side walls.

Figure 7-9 *The bottom of the base disk, showing the guides for positioning on the sundial walls*

Build the Disk Support

Let's fix that front-heavy disk by building a curved support guide for it (Figure 7-10).

Figure 7-10 *By scoring the foam core you can shape it into a curve*

Here are the steps you'll need to follow:

1. Take your 12" × 4.5" piece and score it crosswise every 1/4". Take care to cut through the top layer of paper on the foam core, and slightly into the foam itself, but not all the way through.
2. Measure 1 1/8" in from the center of the long edge of the base disk support foot from Table 7-5.
3. Glue the base disk support guide to the base disk support so that its outer edge just reaches that mark (see Figure 7-11).

Figure 7-11 *The disk support guide semicircle should overlap the disk support foot by 1 1/8" and be centered*

4. Cut the overlapping portion of the semicircle off so that the support foot's edges are flush again, and the foot (with the guide glued on top of it) is once more rectangular.
5. Use the glued-on guide arc to position the scored disk support wall. Center the wall and glue it around the guide. You may need to cut a notch in one end of the wall to allow for the USB jack on the Arduino (Figure 7-12).

Figure 7-12 *The disk support wall glued to the guide and the foot (note the notch on one side for the Arduino's USB jack)*

6. Position your base disk on the base walls and figure out where the curved disk support needs to fit against the front wall of the sundial. It's important that the disk support wall support the base disk but not block the inner track (Figure 7-13).

Figure 7-13 *Don't block the track with the curved support*

7. You'll probably need to nick off the corners of the inner base to allow for the disk support's curve, as in Figure 7-14.

Figure 7-14 *You need to cut the corners off of your lower base to allow for the curve of the support*

8. The disk support base foot is longer than it needs to be to allow for repositioning slightly against the front of the sundial. When you've figured out where the support wall needs to be aligned to support the disk without blocking the track, cut the support foot ends off to match the edges of the sundial. You can see how I've made marks on the foot in Figure 7-15 to note where I need to cut off the ends of the support foot.

Figure 7-15 *The completed disk support; I cut mine to come to a point just because*

Finish Parts and Construction

We are getting close now. To finish the hardware and construction of the sundial, we need to:

1. Build an arm for the azimuth servo that will allow it to move the azimuth ring from underneath as it rotates, through the base disk's incised track.
2. Build the elevation arc by cutting and attaching the elevation ring, and mounting the elevation servo.
3. Cut a gnomon.
4. Wire the "sun" LED.

Build the Azimuth Arm

To build the azimuth arm, follow these steps:

1. Cut the azimuth arm from foam core, as per Table 7-6.
2. Taper from one long side to the other such that the narrow end is 1/4″ wide.
3. Glue, screw, or otherwise attach this arm to your servo's horn (I used a circular horn) such that the tip of the arm is 3 5/8″ from the center of the servo shaft.
4. Put a right-angle bend in a piece of wire 4.25″ from one end.
5. Push the wire through two small pieces of foam core and glue to the arm. Make sure the wire bend is at 3 5/8″ from the center of your servo shaft when the arm is attached (see Figures 7-16 and 7-17).
6. Attach the arm to the servo and then align the base disk on the base. The wire should poke up through the track in the base disk (Figure 7-16).

Figure 7-16 *Alignment of the azimuth arm through the main disks*

Construct the Elevation Arc

To construct the elevation arc, follow these steps:

1. Mark two points directly opposite each other on the elevation ring. Cut the ring about 1/4″ beyond each of these marks to create an arc of slightly over 180°.
2. From the outside, narrow edge, pierce one of these marked spots with wire or a thin metal rod (you may find it easier to pre-poke with an awl).
3. Pierce the same wire through the edge of the azimuth disk, creating a pivot. Bend the wire sharply up at the outside of the elevation ring to secure as in Figure 7-17.

Figure 7-17 *Detail of the elevation arc pivot*

4. Position the sub-micro servo on the azimuth ring so that its shaft points outward 180° opposite from the elevation ring pivot (see Figure 7-18). The top of the shaft should be just flush with the outside of the azimuth ring so that it can serve to rotate one end of the elevation arc. Use hot glue if you're brave, or otherwise secure the servo to the azimuth ring using strong tape like duct tape or try electrical tape.
5. Cut a small, square hole in the azimuth ring next to the elevation servo and run its wires through.
6. Attach the servo's horn at the opposite side of the arc from the pivot (I used a single-pointed horn and attached it with electrical tape) such that it points straight "up" into the arc.

Figure 7-18 *Detail of the opposite end of the elevation arc, the elevation servo and the azimuth ring*

7. Immediately next to the elevation servo, make a hole through the center of the azimuth ring with an awl. The azimuth arm's wire should fit into this hole. This is how the azimuth arm will move the azimuth ring, through the base disk's track. This *due-south alignment hole* (see Figure 7-19) will point, you guessed it, due south when the sundial is in its initial state—that is, directly opposite of the gnomon.

Figure 7-19 *The azimuth arm wire poking up through the southern-aligned hole in the azimuth ring*

Cut a Gnomon

The shape of your gnomon will depend on your distance from the equator. Follow these steps to cut your gnomon (see Figure 7-20):

1. Cut a triangle from foam core. The base should be 3 inches long (the radius of the gnomon disk). The angle from the base should be equivalent to your latitude.
2. For higher latitudes, the larger angle will cause your gnomon to get taller, faster. You may have to cut your gnomon back (make the base shorter) so that the elevation arc can clear the tip of the gnomon. This is fine—the rear angle doesn't need to be square or of a particular length. The only thing that matters is that the gnomon maintains an angle away from the center that equates to your latitude.
3. Now you can fit your gnomon in the slot on the gnomon disk.

Figure 7-20 *My gnomon, cut for the latitude in Cork, Ireland. The angle on the left is approximately 52°. Note my gnomon is not square on the back (right side), and is shorter in the base than 3 inches, so it wouldn't be too tall and interfere with the elevation arc.*

Wire Up the Sun

This is the last hardware step! We need a sun. Here's what you need to do:

1. Push the anode and cathode of your LED through the foam core at the center of the elevation arc from the inside such that the "sun" shines directly down (toward the inside of the arc).

2. Solder or otherwise connect the LED to hookup wires (Figure 7-21).

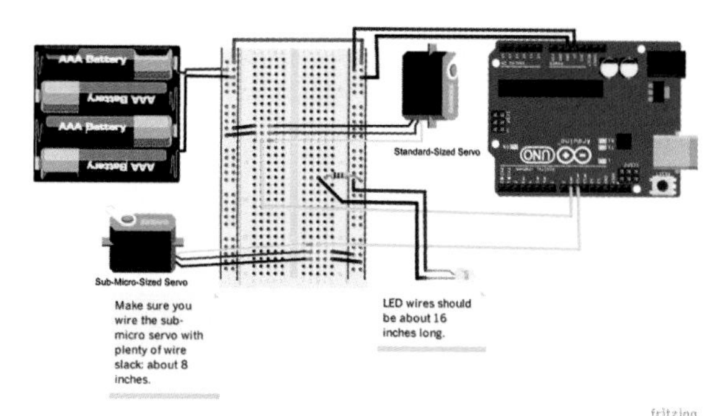

Figure 7-21 *Fritzing diagram with LED wiring*

3. Run both wires down the outside of the arc toward the servo end of the arc. Secure with electrical tape.
4. Poke the wires through to the inside of the arc near the pivot point.
5. Run the wires through the hole cut for the elevation servo wires, leaving enough slack for the elevation arc to pivot comfortably through its range (Figure 7-22).

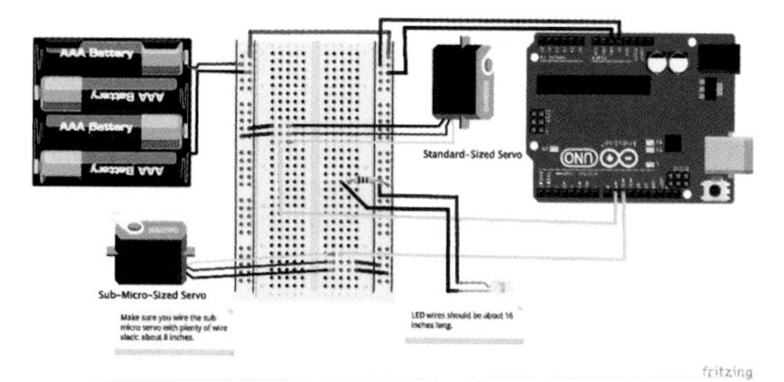

Figure 7-22 *Detail of the elevation arm with the mounted LED and wiring*

Example 7-3 *sundial.js high-level structure*

```
var five = require("johnny-five");
var sunCalc = require("suncalc");

var board = new five.Board();
var servos, sundial;

servos = {
  ❶
```

6. Connect the LED to 5V power through a 220-ohm resistor.
7. Optionally, for appearance, cut two matching arcs in decorative card stock and glue them over each side of the elevation arc to cover up tape and wires.

Code to Make It Go

The physical structure of our sundial is complete. Now for the code. See Example 7-3.

The algorithm for determining the sun's position in the sky given a latitude and longitude and a date and time is not quite rocket science, but it does involve calculating solar mean anomaly, declination, ecliptic longitudes, and a dozen or so other slightly esoteric things. I lost a weekend or two once to delving into the depths of this topic.

However, there's a dead simple way to get the same results: use the `SunCalc` npm package. This package was installed for you when you ran `npm install` earlier, so all we need to do is include it and use it.

Understanding sundial.js

The *sundial.js* script will make our sundial go. It uses `suncalc` to calculate sun position for your latitude and longitude, and positions the azimuth and elevation servos to match that position.

```
};

sundial = {
  ❷
};

function sunPositionInDegrees(date, latitude, longitude) {
  ❸
}

board.on("ready", function() {
  var azimuthServo = new five.Servo(servos.azimuth);
  var elevationServo = new five.Servo(servos.elevation);
  var ticker;

  var tick = function tickTock() {
    ❹
  };

  tick(); ❺

  this.repl.inject({
    ❻
  });
});
```

❶ Configuration options for our servos

❷ Configuration options for our specific sundial, including latitude/longitude

❸ Take SunCalc's sun position in radians and convert to degrees (utility function)

❹ Logic for determining where to position the two servos; moving the servos

❺ Invoking the `tick` function, which is later managed by a `setTimeout`

❻ Adding some stuff to the REPL so we can access it if we want

Set Up Some Configuration in sundial.js

Make a couple of changes to *sundial.js* to customize it for your servos by following these steps:

1. Update the `servos` object.
2. Update the `sundial` object, as shown in Example 7-4.

Example 7-4 *sundial.js*

```
servos = {
  azimuth: {
    pin: 9,
    range: [7, 172], ❶
    isInverted: true,
    center: true
  },
  elevation: {
    pin: 10,
    range: [7, 172], ❷
    isInverted: true,
    center: true
  }
};
```

❶ Update these values with the minimum and maximum angles for your azimuth (standard-sized) servo.

❷ Update these values with the minimum and maximum angles for your elevation (sub-micro-sized) servo.

Next, update the `sundial` object with specifics for your sundial and hardware:

```
sundial = {
  latitude: 45.52,      ❶
  longitude: -122.63, ❷
  tickInterval: 5000, ❸
  msPerDegree: 50       ❹
};
```

❶ Change this value to your latitude. Make sure to use negative if south of the equator.

❷ Change this value to your longitude. Make sure to use negative if west of the Prime Meridian.

❸ How often to check the sundial for updates, in milliseconds. 5 seconds is (much) more than adequate. Adjust if you like; default should be fine.

❹ A ticket to a happy sundial is moving the servos *slowly*. Adjust if you like; default should be fine.

sundial.js Details

Let's look at what else *sundial.js* does. You don't have to make any changes to the chunks of code shown in Example 7-5.

Example 7-5 *sundial.js functionality*

```
board.on('ready', function() {
  var azimuthServo = new five.Servo(servos.azimuth);       ❶
  var elevationServo = new five.Servo(servos.elevation); ❶
  var ticker;

  var tick = function tickTock() {

  };

  tick(); ❷

  this.repl.inject({ ❸
    aServo: azimuthServo,
    eServo: elevationServo,
    tick: tick,
    ticker: ticker
  });
});
```

❶ Servos initialized per config options

❷ Kick of the sundial tick

❸ Make some things available to the REPL

The `tick` function

The `tick` function, shown in Example 7-6, contains the main logic for determining where the servos should be, and moving them to those points. Let's dive in and look at the function in depth. Again, you don't need to make any changes to this code.

Example 7-6 *The entire tick function*

```
var tick = function tickTock() {
  console.log("tick!");
  var position = sunPositionInDegrees( ❶
      new Date(), sundial.latitude, sundial.longitude
    ),
    isFlipped = position.azimuth > 180, ❷
    aPos = (isFlipped) ? position.azimuth - 180 : position.azimuth, ❸
    ePos = (isFlipped) ? 180 - position.elevation : position.elevation,
    aChange = Math.abs(azimuthServo.value - aPos),
    eChange = Math.abs(elevationServo.value - ePos),
    aTime = aChange * sundial.msPerDegree,
    eTime = eChange * sundial.msPerDegree,
    servoTime = (aTime >= eTime) ? aTime : eTime; ❹

  if (ticker) {
    clearTimeout(ticker);
  }

  if (position.elevation < 0) {
    console.log("It is nighttime, silly!"); ❺
    return;
  }
  if (aChange || eChange) { ❻
    azimuthServo.to(aPos, aTime);
    elevationServo.to(ePos, eTime);
  }
  ticker = setTimeout(tick, sundial.tickInterval + servoTime); ❼
};
```

❶ First, `tick` gets the current sun position at our configured latitude and longitude. `sunPositionInDegrees`, shown in Example 7-7, uses SunCalc to obtain the current position in radians, and then converts it to (rounded) degrees.

❷ Recall that we have only 180° of motion from each of our servos (and really not even that). The highest possible elevation for the sun is 90°, when the sun it at the very zenith of the sky. So we can use angles higher than 90° on our elevation servo to fill in gaps for the azimuth servo when the sun is west of 180°.

For example, an azimuth position of 272 with an elevation of 42° is equivalent to azimuth 92, elevation 138. 92 is directly opposite from 272. Our azimuth servo can reach 92, but not 272. 42° elevation reached from the opposite angle to 272 is 138°–90° plus the remaining angle when 42 is subtracted from 90° (or 180° minus 42°). We account for this in `tick` by doing some quick calculations.

❸ When the azimuth is greater than 180, where the azimuth servo cannot reach, we "flip" the whole thing and attempt to reach the position from the opposite azimuth angle. Then we calculate how much change has occurred between the servos' current positions and where they need to be, and from that determine how long the duration should be for the move.

❹ We'll be moving both servos at the same time. Whichever one is going to take longer roughly represents the overall time the servos need to get into position.

❺ It's a sundial, not a moondial!

❻ `Servo.to` takes an optional second `duration` argument in milliseconds. This allows us to move our servos *slowly* and avoid too much torque and strain on our sundial parts.

❼ Cue up another `tick`, making sure to add the `servoTime` to the overall `sundial.tickInterval` so we keep multiple servo moves from happening on top of each other.

Moondials

Yes! Moondials exist! They're somewhat rare and they're usually only accurate during a full moon.

Example 7-7 *sunPositionInDegrees source*

```
function sunPositionInDegrees(date, latitude, longitude) {
  var positionNow = sunCalc.getPosition(date, latitude, longitude);
  return {
    azimuth: Math.round((positionNow.azimuth + Math.PI) * 180 / Math.PI),
    elevation: Math.round(positionNow.altitude * 180 / Math.PI)
  };
}
```

Putting It All Together!

Time to put the pieces together, initialize stuff, and try out the sundial:

1. Pull the elevation servo's wires and the LED wires through the wire hole in the azimuth ring and then through the track slot in the base disk, as seen in Figure 7-23. Rest the azimuth ring in its track on the base disk.
2. Plug the LED and servo wires into the breadboard. Make sure power is hooked up to the azimuth servo as well.
3. Rest the base disk on the rear walls for a moment.
4. Plug in the Arduino and run the following to center both servos:

   ```
   node servos.js
   ```

5. Place the azimuth arm on the azimuth servo such that it points due south (straight toward you).
6. Attach the elevation arc to the servo such that the LED sun is at zenith (straight up/overhead).
7. Guide the azimuth arm through the base disk slot and push through the due-south guide in the azimuth ring.
8. Align the base disk with the base walls using the guides on the bottom of the disk.
9. Position the disk support to support the front of the disk, making sure it does not block any of the base disk's track.

Figure 7-23 *Running the servo and LED wires through the azimuth ring and base disk*

Making It Go!

Now you're (finally) ready to execute the sundial script! Run it and your sundial will slowly move into position:

```
node sundial.js
```

> **Some Things to Keep in Mind**
>
> - Sundials can't adjust for daylight saving time. During daylight saving time, the sundial will run one hour slow.
> - Sundial accuracy varies throughout the year as solar noon "wanders" (see Wikipedia (*http://bit.ly/19LXyFU*)).

Figure 7-24 *Early afternoon in Portland, OR (with a 45° gnomon)*

What's Next?

Here are some additional exercises to try:

- How could you adapt *sundial.js* to work in the southern hemisphere?
- The positioning of the gnomon in our sundial results in shadows that are mostly only on the top (northern) half of the gnomon disk. How else could you position the gnomon to maximize shadow range?
- Could you adjust *sundial.js* to automatically account for daylight saving time?
- Could you adapt the sundial to turn into a moondial in certain instances? (Hint: `SunCalc` also provides moon position data.)

To compute where to put hour lines for your sundial, the equation for the spacing between hour marks is:

$$\hat{I}_{,} = Tan^{-}1(SinL * TanH)$$

*Where **L** is latitude and **H** is the hour.*

8 Spooky Lights

By Anna Gerber

NodeBots and Halloween are two of my favorite things, so with CampJS IV falling on October 31, I found the perfect excuse to build an interactive installation combining the two (see Figure 8-1).

Johnny-Five isn't just useful for programming robots: you can use it to add web interfaces to hardware devices, or you can augment web applications with physical interfaces by connecting sensors and actuators that sense or act upon the physical world. This project demonstrates how easy it is to use Node.js and Johnny-Five to develop application programming interfaces (APIs) for controlling hardware devices.

You'll step through constructing a set of spooky holiday lights using bi-color LED matrices. Then you'll write a web application with a user interface and an API for displaying seasonal patterns and messages on the lights.

Figure 8-1 *Halloween lights with LED matrix faces*

Bill of Materials

To build a string of holiday lights, you'll need the materials listed in Table 8-1.

The quantities for wire and heat shrink tubing are provided as a guide—you can make your string of lights longer or shorter by increasing the length of wires between the lights. You can also vary the project by connecting more or fewer matrices, up to a total of eight to suit your application.

Figure 8-1 shows an example of decorating the lights with silk leaves and miniature jack-o-lanterns, but if Halloween isn't your thing, you can decorate the lights in a theme with your favorite holiday instead. If you have access to a 3D printer, you may wish to 3D print custom enclosures.

Table 8-1 *Materials you will need*

Count	Item	Notes	Estimated price
1	3m hookup wire in red	You're using 13 × 0.12 mm stranded wire, equivalent to AWG 14	$0.50
1	3m hookup wire in black		$0.50
1	3m hookup wire in yellow		$0.50
1	3m hookup wire in green		$0.50
1	3m heat shrink tubing	To fit four hookup wires (e.g., 6 mm)	$5.00
1	64 cm small heat shrink tubing	To fit two hookup wires (e.g., 3 mm)	$1.10
4	Female to female short jumper wires	One each of red, black, yellow, and green	$0.25
7	Four-pin header plug with crimp pins	One for each matrix. Alternatively, you can use extra female to female jumper wires cut in half	$0.60 each
7	HT16K33 8 × 8 LED matrix modules	You can connect up to eight	Between $7.80 and $15.95 each
1	Arduino	You're using an Arduino Nano v 3.0 compatible board with ATmega328	$7.50
1	USB cable for Arduino	Preferably extra long	Comes with Arduino
1	Enclosure	Houses the Arduino	Free (reuse packaging)
	Solder		
	Cable ties		

Table 8-2 *Optional materials for decorating the lights*

Item	Estimated price
Plastic jack-o-lanterns to fit matrices	$2.50
Glue or modeling clay for affixing lanterns	$2.00
Silk leaves	$2.00
Green cable ties for attaching decorative leaves to lights	

Choosing a Controller Board

There are various manufacturers of HT16K33-based I2C 8 × 8 LED matrix boards that come in different sizes and in single-color as well as bi-color versions. We are using bi-color OCROBOT matrices from AliExpress. Adafruit's bi-color square LED Matrix with I2C backpack units provide wonderfully vibrant LED colors and respond to the same I2C commands; however, the red and green colors are inverted on the OCROBOT board.

Background

Before you get started, it's important to understand a little bit about the components you'll be using.

Light-emitting diodes

The D in LED stands for diode. Diodes are polarized semi-conductor components. You can think of them like one-way valves that only operate when electrical current flows through in the correct direction.

Components in electrical circuits typically convert electrical energy into other forms of energy including light, heat, sound, or kinetic energy. When current passes through an LED flowing from the anode (positive lead) through to the cathode (negative lead), the LED emits light.

What's an LED Matrix?

An LED matrix is a component that combines multiple LEDs in a 2D grid arrangement. For this project, you'll be working with 8 × 8 LED matrix components. Each component is arranged as a square-shaped dot matrix of 8 rows by 8 columns, giving a total of 64 LEDs per matrix. Matrices are available with single-color, bi-color (red plus green), or RGB (red, green, and blue) LEDs.

LED matrix components take care of the hard work of wiring together all of the individual LEDs, and make use of common cathode and anode pins to reduce the number of wires you need to connect to communicate with the LEDs. Controlling an 8 × 8 matrix directly from Arduino requires 16 input/output (I/O) pins (one for each row and column).

If you were connecting the Arduino I/O pins directly to an LED matrix, you'd also need to add resistors to the circuit to reduce the current that flows through the LEDs. Current-limiting-resistors are important because if too much current passes through an LED, the excess energy will be converted to heat instead of light, which reduces the efficiency and lifespan of the LED, and can eventually lead to failure of the component. So the wiring of the circuit would start to get pretty messy, and you'd soon run out of pins on your Arduino if you tried to drive several LED matrices directly at the same time.

Fortunately, a simpler solution exists: you can use a driver IC, which reduces the pins required to drive a matrix by multiplexing the display. With multiplexing, patterns are displayed by scanning the matrix to refresh a single row or column at a time. If the refresh rate of the matrix is high enough, the scanning will happen so quickly that the image displayed will appear steady to your eyes due to the persistence of vision effect (the afterimage that remains in your retina).

You'll use an HT16K33 controller driver, which takes care of the multiplexing and communicates with the Arduino via I2C using just two wires.

I2C?

Inter-integrated circuit (I2C) is a serial bus developed in the early 1980s for communication between integrated circuits on computer motherboards. A communications *bus* like this allows data and control signals to be sent to devices connected in parallel. You'll only be connecting LED matrices; however, there are many types of components that use I2C, and you could connect any of them via the same bus. I2C simplifies communication between a microcontroller and connected devices down to just two lines: *SDA* for serial data and *SCL* for serial clock. This makes for a very efficient use of I/O pins: you could continue to add devices to the bus but you'd still only use two I/O pins on the Arduino to communicate with all of them.

Assembling the Lights

Now that you know a bit about how I2C matrices work, you're ready to build your lights! You'll begin by preparing your matrices, and then you'll construct a cable assembly and connect the Arduino. Next, you'll connect each matrix section by section, and finally you'll decorate your lights.

Tools Required

Many of the tools (Figure 8-2) you will need to build the lights can be found around the home or office:

- Wire strippers and cutters
- Pliers
- Scissors
- Hobby knife
- Ruler or measuring tape
- Marker pen
- Soldering iron
- Heat gun or hair dryer

This project does involve soldering; however, you'll mostly just be joining wires, so a basic iron will do. If you don't have wire strippers, you can use scissors or a hobby knife in a pinch. However, you'll be stripping a lot of wires in this project, so do yourself a favor and use a dedicated tool.

You can use a heat gun to quickly and easily shrink the heat shrink tubing, but if you don't have access to one, a hair dryer used on the hot setting will also work.

Preparing the Matrices

Each device connected in an I2C bus needs a unique address (Table 8-3). Addresses are represented using bits that have a binary value of 0 or 1.

Table 8-3 *HT16K33 matrix addresses*

Address	A2	A1	A0
0x70	0	0	0
0x71	0	0	1
0x72	0	1	0
0x73	0	1	1
0x74	1	0	0
0x75	1	0	1
0x76	1	1	0
0x77	1	1	1

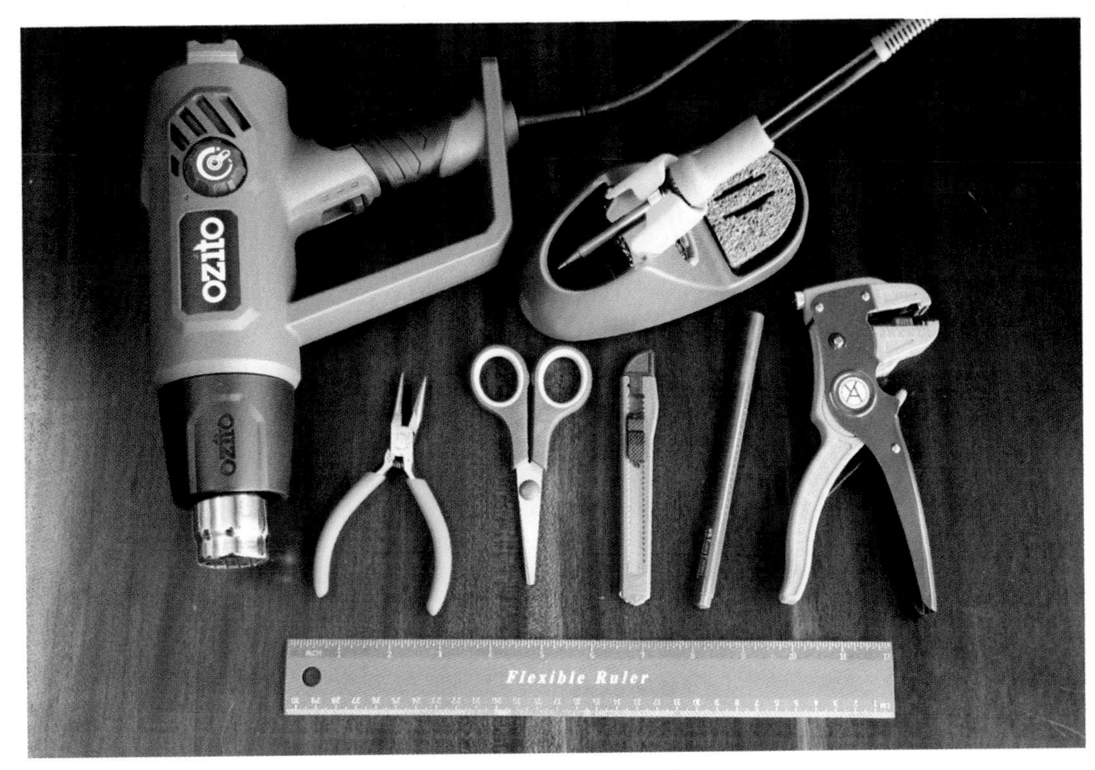

Figure 8-2 *Required tools*

Typically, I2C uses a 7-bit address space, which provides 2^7 (128) addresses, of which 16 are reserved. This means that 112 devices can be connected on a single I2C bus. However, many components implement a more limited range of addresses in practice: your HT16K33 matrix controller board only provides three bits for specifying addresses, so you can connect a maximum of eight matrices. Addresses are assigned by shorting a combination of the three jumpers (A0, A1, and A2) on the back of the board. If you are using single-color Adafruit mini LED backpacks, be aware that you can only connect four of them together, as they have two instead of three jumpers for setting I2C addresses. To prepare the matrices, follow these steps:

1. If your matrix boards require assembly such as soldering the matrix or set of header pins to the controller board, you will need to do that first. Be sure to follow any directions provided by the manufacturer to orient the matrix correctly on the controller board.

2. Set unique addresses for each of your matrices by soldering across the jumpers on the back of the controller board. For example, Soldering A0 and A2 as shown in Figure 8-3, will set the address to 0x75. Assign addresses to the matrices sequentially from 0x70. Holding the soldering iron close to the pads, touch the solder against the tip of the iron to create a small blob of molten solder, and then quickly transfer the blob of solder to the pads.

To avoid heat damage that can cause the pads to lift off the board, minimize the time that the soldering iron is touching the PCB—it should only take about a second for the blob to transfer. Clean the tip of the soldering iron to remove any excess solder before moving on to the next jumper.

3. Use a marker pen to write the address of each matrix along the side of the matrix as you go—this will make it easier to see the address when you are testing the lights later, without having to flip the matrix over to look up the table.

Figure 8-3 *Soldering I2C address jumpers*

Constructing a Cable Assembly

You'll be building a *cable assembly* to connect the matrices together. The cable assembly bundles the four wires used for communication and power. The circuit you'll be using is shown in Figure 8-4. You'll use different colors to identify the I2C lines: data (yellow) and clock (green), as well as the 5V power (red) and ground (black) wires that provide power to the matrix controller boards.

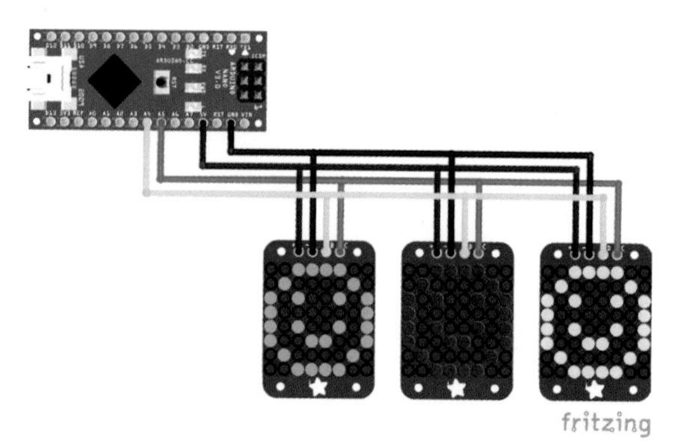

Figure 8-4 *I2C matrix circuit*

Arduino I2C

Arduino boards support I2C communication via fixed pins for data (SDA) and control (SCL).

You're using an Arduino Nano, so the I2C pins are A4 (analog pin 4) for data and A5 for control.

If you decide to use another type of Arduino, the pin numbers may differ. Table 8-4 shows the pin assignments.

Table 8-4 *I2C pins for various Arduino boards*

Board	SDA	SCL
Uno, Nano	A4	A5
Leonardo	2	3
Due	20 or SDA1	21 or SCL1
Mega2560	20	21

You'll construct the cable assembly in sections, one matrix at a time. If you're not very experienced with soldering, this project will give you lots of practice! You'll begin by making lengths of wires to run from the Arduino to the first matrix. You can use female header connectors if you want to have the option of removing the Arduino later, or you could solder directly to the pins if you prefer. Here are the steps you'll need to take:

1. Cut a 34 cm length of each color of hookup wire and strip 2 cm off both ends of each wire.
2. Then, cut four colors of female-female jumper wires in half and strip 2 cm off the cut end.
3. Join a half jumper wire to the matching colored length of hookup wire. An in-line splice is an effective way of joining the wires for this project: join the wires by crossing the two stripped ends together at their centers, and then wrapping each end around the other, one at a time, as in Figure 8-5.
4. Hold the soldering iron against the joined wires to heat them first, then touch the solder to the wires. It should only take a few seconds for the solder to start melting onto the wires. Run solder along the length of the join. Pull the solder away first and then remove the iron about a second later to get a nice finish.
5. Slip a length of small heat shrink tubing about 2 cm long onto the wire and use the heat gun to shrink it to completely cover where you have joined the wires.
6. Repeat this process for all four colors of wire.

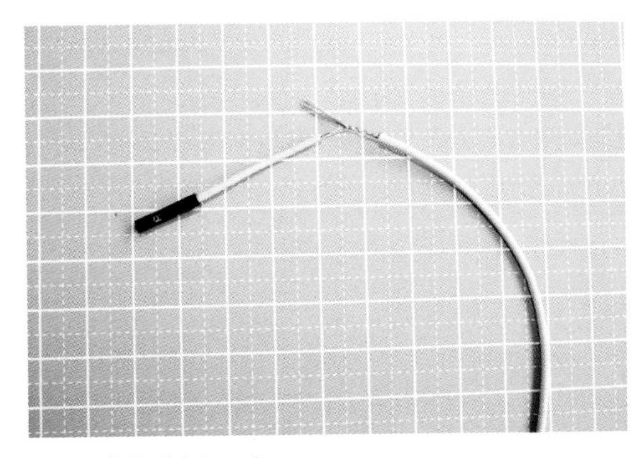

Figure 8-5 *Joining wires*

Connecting the Arduino

Use a small plastic box as an enclosure for the Arduino by cutting holes for the USB cable and matrix cable assembly on either end of the box using a hobby knife. Any small plastic box that fits your Arduino snugly will do the job. Candy containers or plastic packaging from electronics components make great enclosures. Once you're ready, follow these steps:

1. Insert the Arduino into the enclosure, plug in the USB cable, and connect the long header wires you have just made to the pins on the Arduino Nano: red connects to 5V, black to GND, yellow to analog pin 4, and green to analog pin 5 (or I2C pins for your Arduino board).

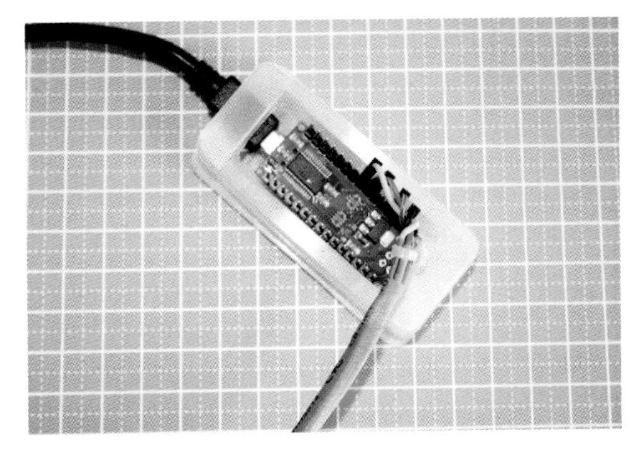

Figure 8-6 *Connecting wires to the Arduino*

2. You'll use green heat shrink tubing as a sheath to bundle the four wires together. This color choice is purely for aesthetic reasons, so that the bundled wires look like a pumpkin stalk. You might like to use a different color of heat shrink tubing, or alternatively, the wires could be held together using cable ties, tape, cable loom, or string. Cut a length of the larger heat shrink tubing approximately 15 cm long.
3. Feed the wires through and shrink the tubing around the wires at the Arduino end, then shut the enclosure and wrap a cable tie around it. The enclosure will

help to prevent the jumper wires from being dislodged as you work on the rest of the cable assembly. About half the length of your wires should be sticking out from the heat shrink; this is deliberate to give us more leeway when joining wires for the next section.

Connecting a matrix

You'll be repeating the steps in this section for each matrix, building up your cable assembly one section at a time.

The wire lengths given in the following will space your lights out approximately 30 cm apart. If you want a different gap between the lights, adjust the longer wires to desired length + 4 cm.

Cut a length of heat shrink tubing approximately 18 cm long (long enough to cover the unbundled length of wires currently sticking out of your heat shrink, plus a few extra centimeters). Thread the heat shrink tubing over the wires that are connected to the Arduino (or previous matrix), but don't shrink the tubing yet.

Next, make a branch jutting off from the main wire to connect to the matrix:

1. Cut a 34 cm length of wire as well as a 4 cm length of hookup wire in each color.
2. Strip 2 cm off each end of the 34 cm wires.
3. Strip 2 cm off one end of the 4 cm wires, and 3 mm off the other end.
4. For each color of wire, create a three-way join to attach the main wires (the 34 cm wires) with the branch wire (the 4 cm wire). Place the short length in parallel with the existing wire so that the stripped wires are pointing in the same direction and twist them together, then use an inline splice to join those wires with the 34 cm length of wire that you just cut, as shown in Figure 8-7.
5. Cover each join in the wires with small heat shrink.
6. As you go, arrange the position of where the wires join so that they will branch off from the main wires in the order of pins on your matrix board when the wires are bundled together. For the OCROBOT boards, this order is green (control), yellow (data), black (ground), red (5V) from left to right when viewing the matrixes from the front, but this order should be reversed for the Adafruit boards.

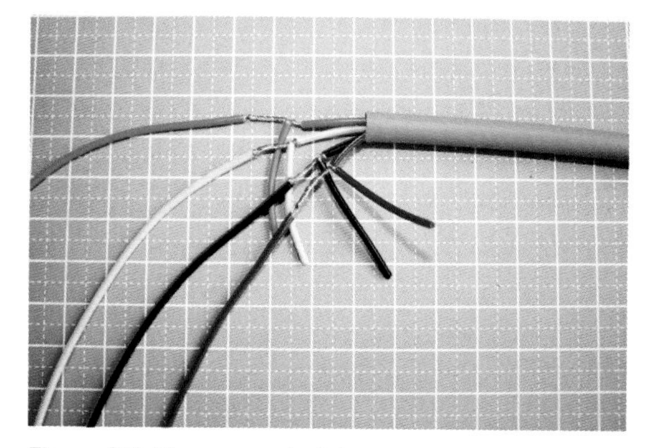

Figure 8-7 *Three-way wire join*

You'll also need to create a three-way join with the larger heat shrink tubing to bundle your wires (you can see what you're aiming for in Figure 8-8):

1. Slice a slot in the bottom of the unshrunk tubing that is already threaded on to the wires. Slide it over the three-way joins so that the tubing completely covers the joins, with the small wires sticking through the slot, while making sure the other end of the tubing over-

laps with the existing heat shrink on the previous section of wires.

2. Cut a 2 cm length of heat shrink tubing and slice two short slots (about 8 mm) on either side of one end. Slide it over the short wires branching off from the main wires, and push it right up to meet and slightly overlap with the main wires with the slotted end sliding under the heat shrink tubing covering the main wires.
3. Cut another 15 cm length of heat shrink tubing and slide it onto the long wires on the other side of the branch, then push it up to the branch, feeding the end of the heat shrink from the other side of the branch inside.
4. At this point, you can shrink all of the heat shrink around the branch. However, if you are not confident with your soldering, you might want to wait until after you have attached the plug and tested the matrix to make it easier to get at the wires again if necessary.

Use pliers to attach the crimps for the 4-pin plug to the ends of the branch wires, like in Figure 8-8:

1. First, use the pliers to squeeze the metal tabs together along the sides of each crimp around the wire. The top set of tabs should be flattened against the bare wires, while the bottom can be against the insulation.
2. Then flatten each set of tabs by squashing them gently with pliers.
3. Once you have crimped all four wires, insert the crimps into the plug in the order matching the pins on your matrix boards, and you are ready to plug a matrix in!

Flattening the crimps can be tricky so you might want to practice with some spare crimps first. If you don't want to use a plug with crimps, you can use female-female jumper wires that have been cut in half in place of the 4 cm length of wires and plug.

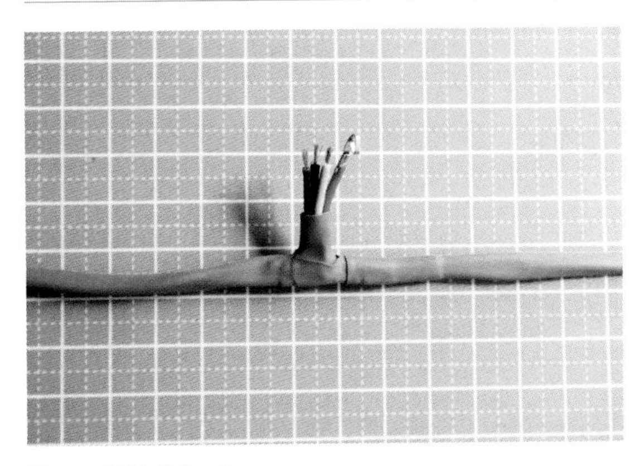

Figure 8-8 *Crimping*

Repeat the steps in this section to connect all of the remaining matrices to the cable assembly. For the last matrix, you don't need to create a branch: add the crimps for the plug directly to the ends of the wires to terminate your cable assembly.

Test as you go!

It's difficult to go back and fix wire joins that are already covered in heat shrink, so check each section of your cable assembly as you go by plugging in a matrix and running the sample program from "Running a Test Program" to see if it works. When testing, make sure none of the exposed ends of the main wires are touching.

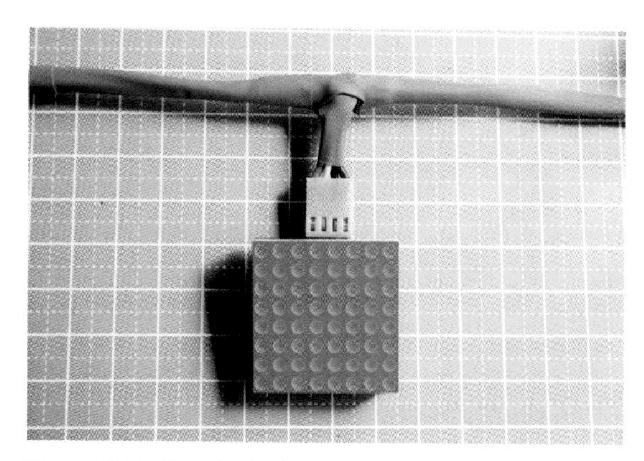

Figure 8-9 *Plugging in the matrix*

Decorating the Lights

Plastic jack-o-lanterns from the local dollar store make perfect enclosures for the LED matrices. Here's what you'll need to do:

1. Hold a matrix inside each lantern and trace around the outline using a marker pen, then carefully cut out the square using a hobby knife. Wipe off any residual marker after the outline has been cut out.
2. Test the size of the hole by pushing the matrix through the hole from inside the lantern so that the LEDs are visible at the front of the lantern. If you cut the outline to fit tightly around the matrix, it should remain in place.
3. If you want the lanterns to be a permanent decoration for your lights, glue along the edges of each matrix to keep the lanterns secure. You can use modeling clay as putty to hold the matrices in place if you want to be able to remove the lanterns to redecorate the lights for other holiday seasons. Roll a snake of clay 6 mm in diameter and press it around the inside edges of the square hole that you cut in the lantern, molding the clay against the sides of the matrix to keep it in place.

Combine yellow and red modeling clay to make orange so that you can match the shade of the jack-o-lanterns—that way, any clay that shows around the edges of the matrix blends in.

4. Use green cable ties from the nursery section of the hardware store to attach silk leaves along the length of the string of lights to complete the look.

Troubleshooting

The main place where things go wrong in this build are the three-way wire joins. If you are not confident in your soldering, use a multi-meter to check continuity on each of your lines. Test the strength of each join by gently tugging on the wires in opposing directions.

When stripping small lengths of hookup wire, sometimes the stranded core at the opposite end of the wire can be exposed when the wire stripper grips and pulls on the insulation. Exposed wires have the potential to touch and cause your circuit to short, so use additional heat shrink or electrical tape to cover them.

Controlling an LED matrix

Let's set everything up to control the matrices.

Preparing the Arduino

You'll need to use at least version 2.4.0 of Firmata when working with HT16K33 devices. In prior versions of Firmata, the size of messages that can be communicated to the Arduino via serial was limited to 32 bytes, which is not large enough for the I2C messages sent to each matrix, however the serial buffer size was increased to 64 bytes from version 2.4.0. At the time of writing, version 2.4.0 of Firmata is available in beta and can be downloaded from the

Firmata sourceforge page (*http://bit.ly/19LYUAq*). See "Arduino" for more details.

All source code for the examples in this book can be found at GitHub (*https://github.com/rwaldron/javascript-robotics*).

Running a Test Program

See the appendix for details on how to install Node.js and Johnny-Five. Once you have Johnny-Five installed, you can run the test program (Example 8-1) to make sure everything is set up correctly.

Example 8-1 *matrix-test.js (Draw a pattern to all matrices)*

```
var five = require("johnny-five");
var board = new five.Board();

board.on("ready", function() {

  var matrix = new five.Led.Matrix({
    devices: 7,
    controller: "HT16K33",
    isBicolor: true
  });

  var heart = [
    "01100110",
    "10011001",
    "10000001",
    "10000001",
    "01000010",
    "00100100",
    "00011000",
    "00000000"
  ];

  matrix.draw(heart);

  this.repl.inject({
    m: matrix,
    heart: heart
  });
});
```

Run the program from the command line as follows:

```
node matrix-test.js
```

You should see a heart displayed on all connected matrices. You can hit Ctrl-D to exit the program.

Matrix Constructor Options

Johnny-Five includes the `Led.Matrix` class for working with matrices. The default device controller for the `Led.Matrix` class is a MAX7219 controller, so you need to set the *controller* option to the constructor to "HT16K33" to work with your I2C matrix boards.

The *devices* option specifies how many matrices you have connected. It is optional—you can leave it out if you are only working with one matrix. This is the simplest way to set up multiple matrices if your devices have been assigned sequential I2C addresses starting at 0x70.

If that is not the case (e.g., if you want to use non-contiguous address ranges), you can provide the *addresses* option to the constructor to specify the exact array of addresses for your matrix boards. If you specify this option you don't need to include the *devices* option. The following code will control a single matrix with the address 0x75:

```
var matrix = new five.Led.Matrix({
  addresses: [0x75],
  controller: "HT16K33",
  isBicolor: true
});
```

The *isBicolor* option indicates whether your controller board is attached to a bi-color or single-color matrix. This is necessary because the data sent via I2C to the matrix boards is handled slightly differently depending on the type of LEDs in the matrix. Leave this option out or set it to false for single-color matrix controller boards such as the Adafruit mini LED Backpack.

Drawing to the Matrix

You can control the LED matrices via the Led.Matrix API. As shown in Example 8-2, you can try these methods out via Johnny-Five's read-eval-print loop (REPL) while the test program is running.

Example 8-2 *Controlling the LED matrices*

```
// controlling all matrices at once
m.draw("M") // draws the letter "M" on all matrices
m.draw(heart) // draw from pattern array on all matrices
m.clear() // clear all matrices

// controlling one at a time
m.device(0).clear() // clear just the 1st matrix
m.device(2).led(0,0,1) // turn on top left LED of 2nd matrix
m.device(0).row(3,255) // turn on all LEDs on 3rd row of 1st matrix
m.device(3).row(1,"00011000") // set pattern for 1st row of 3rd matrix
m.device(0).column(7,0) // turn off all LEDs in last column of 1st matrix
```

Developing a Web Application

You'll use the express framework to develop a web application to control your lights. You'll start by using the express generator to bootstrap your application, then add routes for your API, and finally add a user interface.

The full source code for the web application that you will be developing can be found in the SpookyLights/ folder in the Make: JavaScript Robots repository on GitHub (http://bit.ly/1OebhG4).

Development Tools

Make sure you have Node.js installed and that your Arduino has been flashed with Firmata version 2.4.0 or better, as described in "Preparing the Arduino". You'll also need a text editor or IDE for editing your JavaScript programs (e.g., SublimeText), and a terminal for running commands.

Using the Express Generator

The quickest way to get started with express is to use the express generator. You can install the generator using npm, and then run it from the command line:

```
npm install -g express-generator
express lights-app
```

If you get an error installing `express-generator`*, you can try prefixing the npm command with* `sudo`*, or installing Sudo fix (http://bit.ly/19LYYjT) first.*

This will create a directory called *lights-app/* and generate a complete express web application into that directory. Install the required libraries for the generated app with npm:

```
cd lights-app && npm install
```

To view the web application, run the following command from the terminal and then visit *http://localhost:3000/* in your browser.

```
node bin/www
```

From the command line, install Johnny-Five using npm with the save option so that the latest version will be added as a dependency in your application's *package.json* file:

```
npm install johnny-five --save
```

Example 8-3 *Excerpt of app.js (require johnny-five and set up for Led.Matrix object)*

```
var routes = require("./routes/index");
var lights = require("./routes/lights");
var five = require("johnny-five");
```

```
var board = new five.Board();

board.on("ready",function(){
  var matrices = new five.Led.Matrix({
    devices: 7,
    controller: "HT16K33",
    isBicolor: true
  });
  app.set("matrices", matrices);
});
```

Developing an API

The generated application comes with boilerplate for view templates and routes, and includes commonly used libraries. You'll need to include the Johnny-Five library in app.js and then use the library to create an Led.Matrix object for controlling your matrices. You'll store the object in the express application settings so that you can access it throughout the application as shown in Example 8-3.

You'll also replace the routes that express has generated for users with routes for your lights API. In app.js, do a search on *users* and replace with *lights* and rename *routes/users.js* to *routes/lights.js*. Edit *routes/lights.js* and delete any existing routes, and then add the API routes for the lights API as shown in Example 8-4.

> **Installing on a Single-Board Computer**
>
> At this point, your web application is running locally on a laptop or PC and the lights are connected directly to the computer via USB. This isn't the most convenient setup for a semi-permanent installation. Node.js and Johnny-Five are available on single-board computers like Raspberry Pi, BeagleBone Black, and Intel Galileo, so you could connect your Arduino via USB to one of those and run your web application as is. However, those devices also support I2C, so you could eliminate the Arduino completely and use their on-board GPIO pins directly by utilizing the third-party I/O plug-ins that have been developed for Johnny-Five. For more information on available plug-ins and how to use them, check out the IO Plugins page on the Johnny-Five GitHub site (*http://bit.ly/19LYZEh*).

Example 8-4 *lights.js (Excerpt) (Add routes for API)*

```
// clear all matrixes
router.post("/clear", function(req,res) {
  var matrices = req.app.get("matrices");
  if (matrices) {
    matrices.clear();
    res.send("Cleared all matrices");
  } else {
    res.status(500);
    res.send("Matrices not ready");
  }
});

// draw pattern for a single matrix
router.post("/draw/:device", function(req, res) {
  var device = req.params.device;
  var data = req.body;
  // get johnny-five matrices object from express app
  var matrices = req.app.get("matrices");
  if (matrices) {
    matrices.device(device).draw(data);
    res.send("Updated matrix " + device);
  } else {
    res.status(500);
    res.send("Matrices not ready");
```

```
    }
  });
```

The draw route includes a parameter and device to indicate the index of the matrix device that you want to draw to. You get your Johnny-Five matrices object from the express application settings, and then use the draw method to send the data from the POST request to the matrix. You need to check whether the matrices object exists first, because when the application first starts, there will be a slight delay before the board is ready. This check also allows the app to run and respond to requests if there isn't a board plugged in.

To keep things simple, the sample application only includes API operations for drawing patterns for individual matrices, and for clearing all or individual matrices, but you could add more routes in *routes/lights.js* to correspond to the other methods from johnny-five's Led.Matrix API such as led, row, or column.

Adding a User Interface

The UI for the sample app is in *index.jade* (see Example 8-5), and includes buttons to trigger the API operations that you developed in "Developing an API" and a select element for selecting which matrix device to control.

The sample application also includes a graphical user interface developed using RaphaelJS for drawing patterns to send to the matrices, which you can see in Figure 8-10. You can find the custom `MatrixView` class in *public/javascripts/matrix-view.js* in the source code. Clicking on each square in the grid will cycle through all possible LED colors: green (displayed using web color "Chartreuse"), yellow ("Orange"), red ("OrangeRed"), and off ("Black"). Note that this is hardcoded to match the order of colors used by the OCROBOT boards.

Example 8-5 *index.jade (HTML UI)*

```
h1= title
  div
    button.reset Reset pattern
  div#matrix.matrix
  div
    button.send Send to matrix
  select.device
    option(value="0") 0
    option(value="1") 1
    option(value="2") 2
    option(value="3") 3
    option(value="4") 4
    option(value="5") 5
    option(value="6") 6
  div
    button.clear Clear all
  div.status
```

If you are using an Adafruit board, you should switch the green and red colors in the colors array in MatrixView's constructor. MatrixView's print method serializes the pattern of colors from the grid into an array of strings in the format accepted by Led.Matrix.draw. For example, for the face in Figure 8-10 the data would be:

```
[
  "00000000",
  "02200220",
  "02100120",
  "00000000",
  "00033000",
  "00300300",
  "00033000",
  "00000000"
]
```

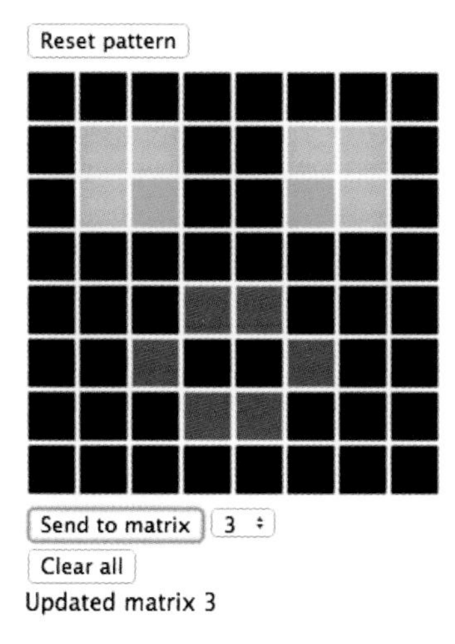

Figure 8-10 *User interface for controlling the matrices*

The code that hooks up the API operations with the buttons is also in *matrix-view.js*. For example, clicking the Send to matrix button triggers a function that gets the device index from the HTML select, gets the LED color data from the MatrixView by calling the print method, and then sends the data via a POST request to your draw API endpoint. Finally, after the POST request is complete, the status div is updated with the message from the XMLHttpRequest response to reflect whether the action was successful (see Example 8-6).

Example 8-6 *Calling the API*

```
$(".send").click(function() {
  var device = $(".device").val();
  $.ajax({
    type: "POST",
    url: "/lights/draw/" + device,
    data: view.print(),
    contentType: "application/json",
    complete: function(xhr, status) {
      $(".status").html(xhr.response);
    }
  });
});
```

Your basic web application is ready to use! Now you can draw 8x8 patterns and send them to any of the connected matrix devices. The sample application has been designed for drawing custom symbols and faces. If you want to display text messages on your lights, you might want to extend the UI to support typing in characters to save having to draw them each time—your draw API already supports any of the types of data that the Led.Matrix draw method allows, including single character strings.

Extending the Application

The application that you've developed provides very basic functionality—it would be nice to be able to save patterns so you could restore them after the lights have been unplugged as well as track pattern history, who created each pattern, and perhaps add access control so that the API isn't open to abuse. Because Johnny-Five runs on Node.js, you can integrate any of the thousands of modules available via npm, to include libraries for working with databases, key-value stores, user authentication, and so on.

What's Next?

I hope that this chapter has given you some ideas for how you could include I2C LED matrix displays in your projects. You could use the techniques described to build a very short cable assembly connecting two I2C matrices to install as eyes inside a jack-o-lantern or interactive toy. And of course you could add a matrix or two your favorite NodeBot, to display status information, or to give your robot facial expressions!

9 CheerfulJ5

By David Resseguie

In this chapter, you are going to build a colored mood light, but instead of expressing the emotion of a single person or object, your CheerfulJ5 project is going to hook into the global CheerLights service: a social experiment that uses Twitter to synchronize the color of lights all around the world (see Figure 9-1). You'll learn to control an RGB LED using Johnny-Five and how to connect to services like ThingSpeak and the Twitter Streaming API to incorporate real-time data from the cloud into your project.

You'll first build and program the circuit using a standard Arduino, then explore how to take it wireless using a Johnny-Five IO Plugin to substitute a Spark WiFi Development Kit in place of the Arduino. However, the use of a Spark is optional and not required to complete the project. You could also substitute alternative hardware platforms using other available Johnny-Five IO plug-ins. To finish up, we'll discuss options for creating an enclosure for your final CheerfulJ5 project to make it suitable for decorative use in the home or office.

Figure 9-1 *Completed CheerfulJ5 project*

Bill of Materials

The electronics for CheerfulJ5 are very simple. The components listed in Table 9-1 can be purchased individually from sites like Adafruit, Sparkfun, Maker Shed, or Amazon. But if you don't already have some of these basics, most retailers carry their own version of a "Getting Started with Arduino" kit that typically contains everything you need. In particular, if you plan to use the Spark WiFi Development Kit to build the wireless version of CheerfulJ5, the *Spark Maker Kit* contains all the components needed to complete the project.

Table 9-1 *Electronic components*

Count	Part	Estimated Price	Part Numbers
1	Arduino Uno R3	$25	MS MKSP99; AF 50; SF DEV-11021
1	Spark Photon or Core	$20-$40	*http://www.spark.io*
1	RGB LED (common cathode or anode)	$2	SF COM-09264; AF 159; AZ B005VMDROS
3	200–300 Ohm resistors	$0.25	SF COM-08377; AZ B00E9Z0OCG
1	Mini breadboard	$4	MS MKKN1-B; AF 65; SF PRT-12043 through PRT-12047
1	Jumper wire kit	$7	SF PRT-00124
1	A-Male to B-Male USB	$3	AZ B000FW60E8; SF CAB-00512
1	3.7V LiPo battery	$9	AF 1578; SF PRT-0034
1	LiPo battery charger	$8	AF 1904; SF PRT-10217

We'll demonstrate a sample enclosure for our CheerfulJ5 project, but this is completely optional. Feel free to make your own creative enclosure or even leave your project exposed and show off the underlying electronics. In our example, you'll create a simple tabletop mood lamp made out of a frosted cylindrical vase that both conceals the electronics and helps to diffuse the light from the LED. Any appropriately sized glass container will work or you can even use a simple piece of tracing or diffusion paper to make your own.

Table 9-2 *Enclosure options*

Part	Estimated Price	Source
Glass vase, globe, or jar	$3–10	Hobby or craft store
Tracing paper or tissue or wax paper (alternative diffusion option)	$3	Hobby or craft store
#216 white diffusion filter (optional, for even better diffusion)	$6	Sold in sheets at theatrical or photography suppliers

Wiring the Circuit

An RGB LED behaves much like a standard LED, except it actually has three LEDs together inside its body: one red, one green, and one blue. You can control the brightness of each of these colored LEDs to mix the colors that you want to produce. There are two types of RGB LEDs: common cathode and common anode. Either type will work for your CheerfulJ5 project; you just have to wire it up appropriately. In our examples, we'll be showing a common cathode LED, but it's easy to modify the circuit to support common anode. The code is essentially the same for either type because Johnny-Five takes care of the translation for you!

The four leads on an RGB LED correspond to red (lead 1), green (lead 3), blue (lead 4), and a common connection (lead 2). Note that the common lead is the second from the flat side of the LED and is the longest of the four. A common cathode RGB LED should have its common lead connected to the ground pin on your Arduino. If using a common anode RGB LED instead, its longest lead should be connected to the 5V pin.

Let's wire up your RGB LED and get ready for a simple test, as shown in Figure 9-2. We show the circuit wired on a mini breadboard so it will be easier to fit inside your enclosure later.

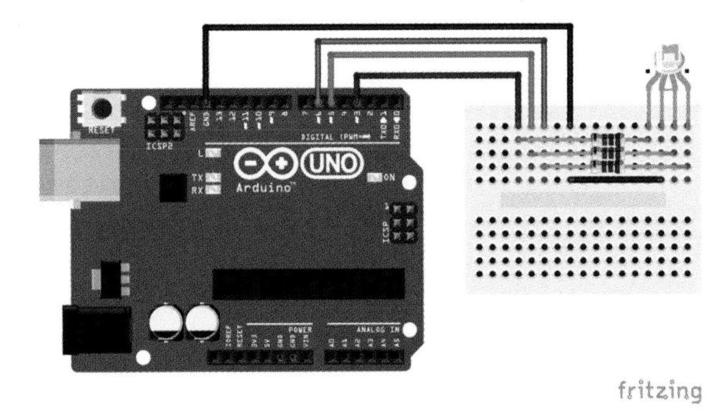

Figure 9-2 *Fritzing diagram for Arduino UNO*

1. Connect leads 1, 3, and 4 of the LED to the Arduino.

Regardless of which type of RGB LED you are using, each individual LED requires its own resistor between it and the Arduino IO pins. You have some flexibility on which value resistor to use, but 200–330 Ohm resistors should be fine. The wiring diagram shown here uses 270 Ohm resistors (red/purple/brown). The RGB LED requires pulse-width modulation (PWM) to control the brightness of each individual LED so be sure to wire it up to supported pins on the Arduino (see Table 9-3).

Table 9-3 *PWM supported pins*

Hardware	PWM pins
Arduino UNO	3, 5, 6, 9, 10, and 11
Spark	D0, D1, A0, A1, A4, A5, A6 and A7

2. Connect lead 2, the LED's common connection, to the Arduino.

   If using a common cathode RGB LED, connect lead 2 to the Arduino ground pin. If using a common anode RGB LED, connect it to the 5V pin.

The CheerfulJ5 code

Now it's time to write some code to control your RGB and connect it to the CheerLights service.

Connecting to the Arduino

Let's get started by creating a bare-bones Johnny-Five application that you can build off of. If you need help installing StandardFirmata, setting up Node.js, or installing Johnny-Five, refer to the appendix.

1. Install Node.js, Johnny-Five, and other dependencies.

If you haven't already done so, install Node.js, followed by the latest version of Johnny-Five from `npm`. You'll also need to install `request`, which you'll use later to query the CheerLights service:

```
npm install johnny-five
npm install request
```

2. Be sure you have the latest version of StandardFirmata running on your Arduino.

   Johnny-Five communicates with the Arduino using the Firmata protocol. StandardFirmata is generally installed by default on most Arduino boards, but can be reinstalled if necessary using the Arduino IDE.

3. Create a new file named *cheerful.js*.

   We'll start with a "Hello World" program that loads the Johnny-Five library, connects to the Arduino via USB, and then prints "Hello World" once the connection is established. Create a new file named *cheerful.js* and enter the code shown in Example 9-1.

Example 9-1 *cheerful.js (bare bones version)*

```
var five = require("johnny-five"),
var board = new five.Board();

board.on("ready", function() {
    console.log("Hello World");
});
```

That's not a lot of code to talk to the Arduino, so what's going on here behind the scenes? When the `five.Board` object is created, Johnny-Five automatically looks for an Arduino connected via USB. Once it's successfully established communication with the Arduino, the `ready` event is triggered. You use the `on("ready")` function to listen for this event and then start executing the main portion of your application. The callback function here is where you will include the bulk of your CheerfulJ5 code later.

4. Let's run a quick test to be sure everything is configured properly.

   Plug the Arduino into the computer using the USB cable and run the following command:

```
node cheerful
```

   The program should print "Hello World" then display a prompt awaiting input from the console. You can just press Ctrl-C a couple times to exit.

You've successfully established a connection to your Arduino. Now on to the fun stuff!

Controlling an RGB

Johnny-Five includes an `Led.RGB` class for controlling an RGB LED. The `Led.RGB` class has many of the same functions as the standard `Led` class, but adds additional functionality for controlling the color. Let's add to your bare-bones application by defining an RGB LED object and setting its color:

1. Initialize an `Led.RGB` object to represent your LED (Example 9-2).

Example 9-2 *cheerful.js (initialize LED)*

```
var five = require("johnny-five");
var board = new five.Board();

board.on("ready", function() {

  var led = new five.Led.RGB({
    pins: {
      red: 3,
      green: 5,
      blue: 6
    }
  });
});
```

This code is the same as the "Hello World" program before, but once the Aruino is connected, we define an RGB LED and tell Johnny-Five that it is connected on pins 3, 5, and 6 for the red, green, and blue leads, respectively.

Alternatively, the `Led.RGB` class specifies a shorthand constructor:

```
var led = new five.Led.RGB([3,5,6]);
```

This version will normalize an array of pins in `[r, g, b]` order to an object that is shaped like:

```
{
  red: r,
  green: g,
  blue: b
}
```

2. If you're using a common anode RGB, you'll need to let Johnny-Five know.

   Using the full constructor, set the `isAnode` property to `true`. After that, all other commands are the same as a common cathode RGB. Johnny-Five translates everything for you automatically!

```
// Initialize a Common Anode RGB LED
var led = new five.Led.RGB({
  pins: {
    red: 3,
    green: 5,
    blue: 6
  },
  isAnode: true
});
```

3. Set the color of the LED.

   To set the color of the RGB LED, you make a call to `Led.RGB.color()` and pass either a hexadecimal color string or an array of the form `[r, g, b]`. For example, if you want the RGB to be red, you can pass the string `"#ff0000"`:

```
led.color("#ff0000");
console.log( led.color() );
```

   Calling `Led.RGB.color()` without a color value causes it to return the current red, green, and blue values. In this case, you simply log the returned object to verify that red is set to its maximum value (255) and the green and blue values are both zero.

4. Run the application again (`node cheerful`) and the RGB should turn red.

Using the Node.js Read-Eval-Print Loop

At this point, it might be helpful to experiment with the available `Led.RGB` functions. The Node.js read-eval-print loop (REPL) provides a way to interactively run JavaScript from the Node.js command prompt. To take advantage of this functionality, you need to inject the `led` variable into the REPL so you can interactively enter Johnny-Five commands to control your LED. Add the following code inside the `board.on("ready")` function:

```
this.repl.inject({
  led: led
});
```

Now you can execute the application again (`node cheerful`) and enter commands at the prompt. For example, to make the LED blink once per second, enter `led.blink(1000)`. Feel free to experiment with other `Led.RGB` functions to better understand what each one does.

Defining the CheerLights Color Map

For convenience sake, let's define a map of color names to hexadecimal values that can be passed to `Led.RGB.color()`. To make your code more modular, put it in its own file that you can `require` in your main application:

1. Create a new file and name it *cheerlights-colors.js*.

   In this file, let's define the color names supported by CheerLights, as shown in Example 9-3.

Example 9-3 *cheerlights-colors.js*

```
var colorMap = {
  red: "#ff0000",
  green: "#00ff00",
  blue: "#0000ff",
  cyan: "#00ffff",
  white: "#ffffff",
  warmwhite: "#fdf5e6",
  purple: "#a020f0",
  magenta: "#ff00ff",
  pink: "#ff69b4",
  yellow: "#ffff00",
  orange: "#ff8c00"
};
module.exports = colorMap;
```

You might have to tweak the color definitions to more accurately represent each color for your RGB LED and enclosure.

2. Update *cheerful.js* to use the CheerLights color definitions.

   Now, instead of setting the color directly by passing in a hexadecimal value, update your previous code to reference one of the standard color names from the color map object. The only changes are to `require` the new *cheerlights-colors.js* at the beginning of your application and to update the call to `Led.RGB.color()`:

   ```
   var colorMap = require
     ("./cheerlights-colors");
   // ...
   led.color(colorMap["red"]);
   ```

Now that you have your RGB LED set up and are able to control its color, it's time to hook up your application to CheerLights!

Accessing the CheerLights ThingSpeak API

The CheerLights service (*http://www.cheerlights.com*) synchronizes color via Twitter messages sent to @cheerlights (*https://twitter.com/cheerlights*) or using the hashtag #cheerlights (*https://twitter.com/hashtag/CheerLights?src=hash*). CheerLights clients around the world can listen for updated color commands and update their own color accordingly. We are going to explore two methods for listening for new color commands. First, we'll look at how to query the CheerLights ThingSpeak API. Then, we'll see how you can tap into the Twitter Stream directly.

ThingSpeak is an open source platform for the Internet of Things with a RESTful API for querying historical and real-time data. To simplify access to the CheerLights service, the creators of the project created a CheerLights ThingSpeak "channel" that captures and stores all incoming color messages. The CheerLights project utilizes the TweetControl App from ThingSpeak to listen to Twitter for the keyword "cheerlights" and update the CheerLights ThingSpeak channel to store the latest requested color. The ThingSpeak Channel API can be used to query for the current color so you can set your RGB LED's color appropriately.

For CheerfulJ5, we only care about the current CheerLights color, so you need to query ThingSpeak for the last requested color value. ThingSpeak provides a `last` endpoint as part of their Channel API that returns the most recent value for a ThingSpeak channel. The response is availabe as XML, JSON, and plain text.

A call to the `last` endpoint is of the following format: *https://api.thingspeak.com/channels/1417/feed/last.json*.

You can test the Channel API by entering the RESTful API URL directly in your browser. In this case, `1417` is the ThingSpeak Channel ID for the CheerLights stream, so you can request the latest CheerLights color in JSON format (*http://bit.ly/1bQNb5p*). The only value you care about in the returned JSON object is `field1`, which contains the last color string requested.

Let's update your *cheerful.js* code to call the ThingSpeak API, parse out the returned value, and update the color of your RGB LED accordingly:

1. Add a function to *cheerful.js* that calls the ThingSpeak API.

   You'll need to define a new function called `getLatestColor()`, shown in Example 9-4, that asynchronously calls the ThingSpeak API and passes the returned color to a callback function. You will use `request`, a simplified Node HTTP request client, to make your calls to ThingSpeak. You first pass `request` the URL to the `last` endpoint of the ThingSpeak Channel API and tell it that you are expecting a JSON response. The second argument to `request` is a function that is called when your API call finishes successfully. Here you extract the latest color string from `field1` and return it to the `getLastColor()` callback function.

Example 9-4 *The getLatestColor function*

```
function getLatestColor(callback) {
  request({
    url: "https://api.thingspeak.com/channels/1417/feed/last.json",
    json: true
  }, function(error, response, body) {
    if (!error && response.statusCode === 200) {
      var color = body.field1;
      callback(null, color);
    } else {
      callback(error, null);
    }
  });
}
```

2. Update LED color based on returned ThingSpeak value.

   To update your CheerfulJ5 RGB LED, you just need to poll ThingSpeak at a given interval and call `led.color()` with the returned result. Note that you're only asking for the latest CheerLights color, so it's possible that you will miss colors between calls to the API—but that's OK for these purposes. You'll use a standard JavaScript `setInterval()` call to check the color every 3 seconds. You'll also save a reference to the current color so you can output to the console when it changes.

Putting everything together, Example 9-5 shows the complete *cheerful.js* code.

Example 9-5 *cheerful.js (using ThingSpeak API)*

```
var request = require("request");
var five = require("johnny-five");
var colorMap = require("./cheerlights-colors");
var board = new five.Board();

board.on("ready", function() {
  console.log("Connected");
```

```
  var lastColor = "white";
  var led = new five.Led.RGB([3, 5, 6]);

  this.repl.inject({
    led: led
  });

  led.color(colorMap[lastColor]);

  setInterval(function() {
    getLatestColor(function(err, color) {
      if (!err && colorMap[color]) {
        if (color != lastColor) {
          lastColor = color;
          console.log("Changing to " + color);
          led.color(colorMap[color]);
        }
      }
    });
  }, 3000);

});

function getLatestColor(callback) {
  request({
    url: "https://api.thingspeak.com/channels/1417/feed/last.json",
    json: true
  }, function(error, response, body) {
    if (!error && response.statusCode === 200) {
      var color = body.field1;
      callback(null, color);
    } else {
      callback(error, null);
    }
  });
}
```

Try running the application (`node cheerful`) and sending color commands to *@cheerlights* using Twitter to watch your RGB update in sync with your Twitter messages! For example, you might tweet:

```
@cheerlights Let's celebrate JavaScript
by turning our @CheerfulJ5 yellow!
```

The CheerLights server listens for Twitter mentions and parses the color name from the message. CheerfulJ5 picks up the change and turns the LED yellow the next time it checks in with the CheerLights server (assuming another user doesn't change the color again before your application polls the API). See Figure 9-3.

Figure 9-3 *CheerfulJ5 running on Arduino Uno after a blue command was sent to CheerLights*

Using the Twitter Streaming API

Our previous approach to getting the current CheerLights color was to poll the preprocessed ThingSpeak channel. This provides a nice, simple solution, but has a few drawbacks. The constant polling of the ThingSpeak channel is inefficient if there are no updates coming in. You could reduce the polling frequency, but that leads to longer response times once a new color command is sent. For some applications, you may also want to avoid missed colors.

As an alternative, you could instead monitor the Twitter stream directly using the Twitter Streaming API. This requires a little more work to essentially duplicate the functionality of the ThingSpeak CheerLights API, but it means you can get the updates immediately and not miss any colors.

To use the Twitter Streaming API, you need to set up a Twitter Developer account (free) and configure a CheerfulJ5 application:

1. Create a Twitter Developer account (free).

   Go to *https://dev.twitter.com* and log in with your Twitter username and password. If you do not yet have a Twitter account, click the "Sign up now" link under the login form.

   If you have not yet used the Twitter developer site, you'll be prompted to authorize the site to use your account. Click "Authorize App" to continue.

2. Create a new Twitter Application.

   Visit the Twitter Application Manager (*http://apps.twitter.com*) and click the "Create New App" button to get started. When defining a new application, you are requested to enter a name (e.g., "CheerfulJ5"), description, and a website. (You can enter a placeholder URL if you don't have a public website for your project.)

3. Generate Twitter access credentials for your application.

   You need two key pairs to access the Twitter Streaming API: the "Consumer Key/Consumer Secret" and "Access Token/Access Token Secret" for your account. These are found under the "Keys and Access Tokens" tab in the Twitter Application Manager. The Consumer Key (API key) and Consumer Secret (API secret) are automatically generated for us and are listed under "Application Settings." To generate the Access Token, you need to click the "Create my access token" button at the bottom of the page. After doing so, the newly generated Access Token and Access Token Secret are displayed.

4. Save access credentials as environment variables.

   It's best not to keep your access credentials for Twitter in your source code, so let's store them as environment variables that you can access via Node.js.

   Create a file in your home directory called *.twitterrc* that contains your Twitter API credentials:

   ```
   export TWITTER_API_KEY="your API key"
   export TWITTER_API_SECRET="your API se
   cret"
   export TWITTER_TOKEN="your access to
   ken"
   export TWITTER_TOKEN_SECRET="your ac
   cess token secret"
   ```

   To load your credentials automatically, you can add the following to your dotrc file of choice:

   ```
   source ~/.twitterrc
   ```

5. Install `node-tweet-stream` module.

   There are a number of `npm` modules that allow you to easily use the Twitter APIs. There's no need to reinvent the wheel, so let's take advantage of one of the existing solutions. Because you are only interested in the Twitter Streaming API for this project, the `node-tweet-stream` module will suffice for making your queries. You'll just need to install it via `npm`:

```
npm install node-tweet-stream
```

6. Replace the ThingSpeak API call with `node-tweet-stream`.

   The `node-tweet-stream` module is very straightforward to use. You create a new `twit` object and pass in the keys you generated previously. You then tell `twit` what phrase(s) you want to monitor using the `track()` function. Any time a matching tweet is found, an event is fired and you can capture the message using the `on("tweet")` function.

   To use the Twitter Streaming API in place of the ThingSpeak API call, you remove the `getLatestColor()` function and associated `setInterval` call. Instead, you define your `twit` object, tell it to track both "@cheerlights" and "#cheerlights", and update the RGB color when the `on("tweet")` function is triggered.

   Because you're not using the preprocessed data from ThingSpeak, you have to parse through the tweet text and extract the color string manually. Inside the `on("tweet")` function, you'll need to look for an instance of a supported color string in the body of the Twitter message. To do this, extract the keys from your `colorMap` object (representing valid CheerLights colors) as a JavaScript Array. Then use the `Array.prototype.some()` function to iterate through each color string and check to see if it is found in the Twitter message using `tweet.text.indexOf(color)`. Once a match is found, you set the RGB color with a call to `led.color( colorMap[color] )` as before.

Example 9-6 shows the full modified version.

Example 9-6 *cheerful-twit.js (using Twitter Streaming API)*

```
var request = require("request");
var five = require("johnny-five");
var twit = require("node-tweet-stream");
var colorMap = require("./cheerlights-colors");
var board = new five.Board();

board.on("ready", function() {

  var lastColor = "white";
  var led = new five.Led.RGB([3, 5, 6]);

  led.color(colorMap[lastColor]);

  t = new twit({
    consumer_key: process.env.TWITTER_API_KEY,
    consumer_secret: process.env.TWITTER_API_SECRET,
    token: process.env.TWITTER_TOKEN,
    token_secret: process.env.TWITTER_TOKEN_SECRET
  });

  t.track("@cheerlights");
```

```
    t.track("#cheerlights");

    t.on("tweet", function(tweet) {
      // grab a matching supported color in the tweet
      Object.keys(colorMap).some(function(color) {
        if (tweet.text.indexOf(color) >= 0) {
          if (color != lastColor) {
            lastColor = color;
            console.log("Changing to " + color);
            led.color(colorMap[color]);
          }
          return true;
        } else {
          return false;
        }
      });
    });

    t.on("error", function(err) {
      console.log("Error with Twitter stream: %o", err);
    });

  });
```

Now your CheerfulJ5 updates its color in real time without a delay for polling! At this point, the CheerfulJ5 project is complete if you don't plan to make use of a Spark WiFi device. If that's the case, you can skip straight to the end of this chapter for suggestions on packaging the final product.

Going Wireless with the Spark WiFi Development Kit

Let's remove the need to tether your CheerfulJ5 project to your computer by replacing the Arduino with either a Spark Core or Spark Photon WiFi Development Kit.

If you haven't already done so, you'll need to claim your Spark device and configure it for use with Johnny-Five. Full instructions for setting up your Spark device are available in the Appendix. Our code examples here will assume that you have installed the custom VoodooSpark firmware and that Spark credentials have been saved as environment variables as suggested in the appendix.

Adding the Spark to Your Circuit

The wiring for the Spark version of CheerfulJ5 is very similar to what you used before with the standard Arduino. But with the Spark, you can take advantage of the small form factor and fit the entire circuit on a single mini breadboard. This is very convenient when you look at putting it in an enclosure. For simplicity, you can just power the Spark from your computer (via USB) during testing and then switch to an appropriate battery later depending on which enclosure you use. But if you want to go ahead and hook up your battery, the wiring diagram here also demonstrates how you can optionally power the Spark via the `vin` pin (see Figure 9-4).

Now your Spark is ready to go. Next up is to communicate with it in Johnny-Five using the Spark-io IO Plugin.

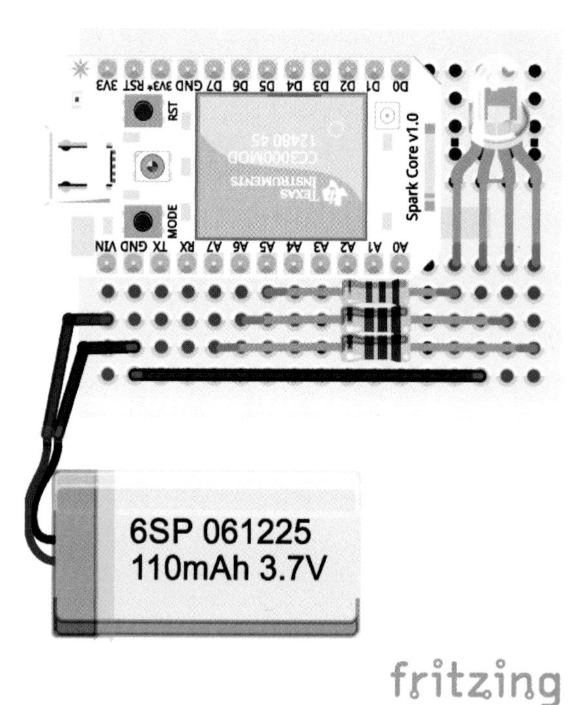

Figure 9-4 *Fritzing diagram for Spark Core*

Using the Spark-io IO Plugin

The Spark-io plug-in is a Firmata-compatible interface that allows Johnny-Five to communicate with a Spark device in the same way it would a standard Arduino. Spark-io takes care of the interface with both the Spark Cloud and VoodooSpark for us. It queries the Spark Cloud to get your device's IP address and port. Then it uses those to connect to VoodooSpark and issue commands to the device. Spark-io can be used standalone or with Johnny-Five. When used alone, the device's pins can be accessed directly via commands like `this.digitalWrite()`. But when paired with Johnny-Five, you get the full power and functionality of Johnny-Five on the Spark!

Using the Spark-io IO Plugin is simple. Instead of using the default `five.Board()` constructor, you use an expanded constructor and pass a reference to a Spark object (along with your Spark credentials) using the `io` property. Once you've created the board using Spark-io, the rest of the code is the same as your previous example using a standard Arduino!

If you don't have a Spark device but would like to use an alternative IO Plugin to connect another Firmata compatible device, the process is essentially the same. You just need to require the necessary plug-in and create the `five.Board` with the `io` property configured as required for your hardware platform. Ensure that you are using I/O pins that support PWM and the rest of the code should work without modification.

For more information on using and creating IO Plugins, visit the Johnny-Five wiki (http://bit.ly/1bQNpcl).

Let's update your previous *cheerful.js* application:

1. Start by copying your existing *cheerful.js* code into a new file named *cheerful-spark.js*.
2. Require `spark-io` to include the IO Plugin in your application:

   ```
   var Spark = require("spark-io");
   ```

3. Replace the call to `new five.Board()` with an expanded Spark-io constructor:

   ```
   var board = new five.Board({
     io: new Spark({
       token: process.env.SPARK_TOKEN,
       deviceId:
       process.env.SPARK_DEVICE_ID
     })
   });
   ```

4. Change the pin assignments for your RGB LED to use PWM pins on the Spark and specify the pins with an `A` prefix:

```
var led = new five.Led.RGB(["A5",
"A6", "A7"]);
```

And you are done! Everything else is the same as before.

Switching to Battery Power

At this point, you have a choice to make. You'll need to switch to battery power if you want to go completely wireless. You have several options. You can power the Spark using a mini-USB backup battery and the included USB port. This is an easy solution, but may be a bit bulky for your intended enclosure later. Alternatively, you can supply power via the Spark's `vin` pin. The Spark datasheet lists an input voltage range of 3.6V to 6.0V. Ideal sources of power can be a 3.7V LiPo battery or 4AA battery pack. Simply connect the positive wire from the battery to the Spark `vin` pin and the negative wire to the Spark `ground` pin (see Figure 9-5).

Figure 9-5 *CheerfulJ5 running wirelessly on a Spark Core*

Packaging It Up

Now that you have a working CheerLights client, let's finish the project by packaging it up in an attractive way. The CheerfulJ5 circuit is pretty simple and lends itself well to a variety of fun applications. This is your opportunity to be creative! Don't feel like you have to do it exactly as suggested here. We'll discuss a few options that will hopefully inspire your own ideas to make CheerfulJ5 unique to you. The examples here all assume you are working with the Spark implementation of CheerfulJ5 because it has a smaller form factor. But if you don't have a Spark, don't let that stop you! You can simply find a similar, but larger, enclosure to use. Or alternatively, you could embed just the RGB LED circuit in your enclosure and run a set of extension wires to an externally placed Arduino. Here are the steps you should follow:

1. First, select a container for CheerfulJ5. One option is to use a very simple enclosure to create a small tabletop mood lamp out of your project by placing your CheerfulJ5 electronics inside a frosted cylindrical vase, as shown back in Figure 9-1. The vase shown here was purchased at a local craft store. It is approximately 5" in diameter by 7" tall and works nicely to conceal the electronics, but also allows for the RGB LED to fully light the lamp.

   If possible, take your CheerfulJ5 project with you to the store so you can make sure you buy a container that is large enough to completely hold the Spark, mini breadboard, and battery. For the best results, power up your RGB LED while you're there to see how it looks inside the container. Experiment with several options, such as frosted or patterned glass, to see how your project will look.

2. Optionally, you may choose to add diffusing material. If you use a clear glass container or would prefer more uniform lighting, you may want to add an insert made from tracing paper, wax paper, tissue paper, or a diffusion filter. Cut the paper or filter to size, depend-

ing on the dimensions of the container, and roll it into a tube. After placing it inside the glass, it should naturally unroll to fill the container. If needed, a small piece of tape can be added to help it maintain its form. If a circle of paper is cut to cover the bottom of your container, it will help reflect light back up as well.

3. Finally, cover the Spark's built-in RGB. The main onboard RGB on the Spark (the board's status light) is a bit bright and may overpower the RGB you're synchronizing with CheerLights. If that's the case for your enclosure, you can fold a small piece of paper over the RGB to block it.

Don't be afraid to think outside the box when packaging your project. You will often find inspiration in unexpected departments within your local craft store. For example, you may try CheerfulJ5 using the following alternative enclosures:

- Small plastic craft container
- Candle holder and wall sconce
- Touch light (disassembled with CheerfulJ5 in place of the standard bulb, using the light's existing batteries and switch)
- Mounted behind a glass Christmas tree ornament
- Small luminary bag

What's Next?

Now it's time to place your CheerfulJ5 in a prominent place and enjoy watching the color change in sync with other CheerLights clients around the world. Because this project allows for a lot of creativity in the final packaging, we'd love to see your final project. Post your photos to Twitter and tag @CheerfulJ5 (*https://twitter.com/cheerfulj5*).

So where do we go from here? Hopefully you'll find the ability to integrate RGB LEDs very useful for future NodeBots projects. LEDs are a great way to provide feedback to users, communicate emotion in robotics, or just add flair to your creation. Connecting to real-time data sources in the cloud like ThingSpeak and Twitter also opens up lots of opportunities for smart and interactive robotics applications. My hope is that CheerfulJ5 has inspired you to build something amazing!

Interactive RGB LED Display with BeagleBone Black

10

By Kassandra Perch

I've always loved playing with colors. Finger-painting, drawing, working with polymer clay: I love working with mixing colors and seeing what happens. So when I got into robotics, I was naturally drawn to individually addressable RGB LEDs: millions of colors and combinations were unlocked for me.

But I wanted a way to interact with these RGB LED chains that used my favorite language, JavaScript. This project shows you how you can use a BeagleBone Black (BBB) and a few lines of code to play with RGB LEDs yourself. This includes using the npm packages open-pixel-control, Johnny-Five, and Beaglebone-io, and some sensors, to create an interactive light display that changes color and intensity based on the environment (see Figure 10-1).

Figure 10-1 *A version of the completed project on display at a conference*

Bill of Materials

Table 10-1 *Bill of materials*

Count	Part	Part number/source	Estimated price
1	BeagleBone Black	MS MKCCE4, AF 1996, SF DEV-12857	$45
1	Wifi USB Adapter (optional)	AZ B003MTTJOY	$10
1	5V power supply (2–10 Amps)	SparkFun or Adafruit	$6–$25
1	Set of RGB LEDs	Adafruit or *http://rgb-123.com*	$6+
1	Pushbutton	Adafruit, electronics store	$1
1	Triple-axis accelerometer	AF 163, SF SEN-09269	$15
1	Photoresistor	SparkFun or Adafruit	$1
1	Half-size breadboard	MS MKKN2; AF 64; SF PRT-09567	$5
1	1k ohm resistor	Electronics shop, online	$0.25

BeagleBone Black

A BeagleBone Black (*http://beagleboard.org*) is a great place to start when working with RGB LEDs in JavaScript. If you've never worked with one, let me give you the quick version: the BeagleBone Black is a small computer, usually running a Linux distro. They're fairly cheap, and easy to obtain. The difference between the BeagleBone Black and the computer on your desk are the GPIO pins available to you—this is what you'll use to control your RGB LEDs, as well as accept input from sensors, buttons, and so on.

The current model at the time of writing is the Rev C. This model is fine to use for these projects, and unless otherwise noted, is what these directions address. However, any model of BeagleBone Black can be usable to complete this project.

The default operating system is assumed for this build, whether that's Debian on the Rev C or Angstrom on the older models.

If you use a different operating system, the software we install may not function correctly.

WiFi USB Adapter (optional)

If you can't be near an Ethernet port every time you want to download or update, a WiFi USB adapter is really helpful. There are many available at stores like Adafruit and SparkFun, and they include the installation instructions.

For configuring your new WiFi adapter, follow these instructions (*http://bit.ly/1bQNJrT*) for Rev C, and these instructions (*http://bit.ly/1bQNGMS*) for older models.

Otherwise, you will need a wired Internet connection and an Ethernet cable.

External 5V Power Supply (Semi-Optional)

These power supplies have a barrel connector and plug into a wall socket:

You need get a power supply that provides at least 1A of current. Be very careful to use a 5V supply 2.1mm inner diameter barrel, center pole positive. If you get the voltage and polarity wrong, you could ruin your board! See

BeagleBoard (*http://bit.ly/1bQNAVz*) for recommended power supplies and other peripherals.

The reason I say this is semi-optional is because the LEDs can function off a 3.3V USB supply, but *not* for chains of more that 20 or so without experiencing some color distortion. So if you plan on using even medium-length chains, you'll want a 5V supply plugged either into the BBB or into the LEDs directly: Adafruit has a great tutorial (*http://bit.ly/1bQNzRe*) for directly powering your RGB LEDs.

RGB LEDs

I currently use two different brands of addressable LEDs for my projects: either Adafruit Neopixels (*http://www.adafruit.com/category/168*) or RGB-123 panels (*http://rgb-123.com*). However, any addressable LEDs that use a WS2812 should work fine.

You can get Neopixels in several places: Adafruit, Maker Shed, SparkFun, and so on. RGB-123 panels can be bought from the manufacturer's website at *http://rgb-123.com*.

Sensors

Cycling through colors is fun, but having your RGB LEDs react to input is even better! Try as many sensor combos as you like, but in this chapter we will use:

- A button
- An accelerometer (three-axis)
- A photoresistor

Miscellaneous

We will use the following items, which happen to be useful for many other electronics projects as well:

- Breadboard
- 1k and 330 ohm resistors (for sensors)
- Breadboard wires

Getting Ready: Software

To start the project, you'll install some software on your BeagleBone Black that will let you more easily communicate with your RGB LEDs.

LEDScape

LEDScape is the software that controls the RGB LEDs on the BeagleBone Black's IO pins. It does this using *Open Pixel Control* (*OPC*), which is a protocol for sending color data over IP. You're going to use an OPC-compliant server on the BeagleBone Black to receive data for the pixels, which you'll send using Node.js.

You're actually going to use a fork of the original OPC designed with these LEDs. It can be found on GitHub (*http://bit.ly/1bQNRHY*). You'll need to follow these steps:

1. Download LEDscape.

   First, SSH into, or log into, your BeagleBone Black. Then, in your home directory, run the commands shown in Example 10-1.

   Example 10-1 *Cloning LEDscape*

   ```
   cd ~/
   git clone https://github.com/Yona-
   Appletree/LEDscape.git
   cd LEDscape
   ```

2. Move a boot configuration file needed by LEDscape into the right folder.

   This file is called a device map, and it allows LEDscape to use pins it otherwise could not.

   If you're on a Rev C BeagleBone, run the commands shown in Example 10-2 within your *LEDscape/* folder. Otherwise, use the commands shown in Example 10-3.

Example 10-2 *Preparing device map on a Rev C BeagleBone Black*

```
cp /boot/uboot/dtbs/am335x-
boneblack.dtb{,.preledscape_bk}
cp am335x-boneblack.dtb /boot/
uboot/dtbs/
```

Example 10-3 *Preparing device map on other BeagleBone Black models*

```
cp /boot/am335x-
boneblack.dtb{,.preledscape_bk}
cp am335x-boneblack.dtb /boot/
```

3. Now, load the modules, and reboot:

   ```
   modprobe uio_pruss
   reboot
   ```

   Your BeagleBone Black should reboot, and you'll either need to SSH back in or wait for the terminal to pop back up on your display.

4. Compile/install LEDscape.

   In your home directory, after your BeagleBone Black restarts and you have logged/SSH'd back in, run these commands:

   ```
   cd LEDscape
   make
   ```

   This will take a few minutes, and output a lot of text. Unless you see an error code at the end, this is normal and means LEDscape is installing properly.

That's it! If there are no errors, you now have the ability to use LEDscape. Before we enable it as a service that runs on boot, we're going to test it.

Wiring Your LEDs

If your RGB LEDs have two wires leading to a connector that are black and white, and a separate black and red wire, never fear! You just have two ground wires. Unless you are using external power, as noted in "Powering Many LEDs", just wire one ground wire from the RGB LEDs into your circuit. If you are using external power, wire the ground and signal wire from the connector to the breadboard/BeagleBone Black.

Wiring Up Your RGB LEDs

Next, we're going to test our software installation by wiring up our RGB LEDs and running a demo of the software.

1. Find your signal pin(s) using pinmap.

   Run the following command inside your *LEDscape/* folder:

   ```
   node pinmap.js
   ```

   This command should output a map of the BeagleBone Black. We're looking for strip index 0 on the old mapping system. As of the time of this writing, this maps to pin P9_22. So wire the signal pin of your strip to whichever shows up on *pinmap.js*'s output.

2. Wire up your RGB LEDs.

   Grab your RGB LED chain and wire all three wires to a piece of breadboard (if you have four wires, see the previous note). Then, wire ground (usually black or white, connects to GND on the LED strip) to P9_1 or P9_2, signal (usually green, connects to DIN on LED strip) to channel 0 as found in *pinmap.js* (pin P9_22 at the time of this writing), and power (usually red, connects to 5V on LED strip) to P9_5 or P9_6.

   If you are using multiple strips, wire the signal pins of those strips to index 1, 2, 3, and so on, up to 47, according to the output from the previous command.

3. Test your LEDscape install and wiring.

   Now you're going to test the software installation and our hardware wiring.

Run the following in your *LEDscape/* folder:

```
./run-ledscape
```

Because there is demo code that is set to run by default, you should see rainbows! Exciting! But you'll notice you can't use your terminal while the server is running. You'll fix this by creating a service that runs the server in the background when the BeagleBone Black is booted up.

4. Set up LEDscape service configuration.

   Before you set up the service to run, you need to make sure it's configured to suit your needs. In the *LEDscape/* folder, run the following:

```
nano ./run-ledscape
```

Look for the line that says something like:

```
./opc-server -p 7890 -c 176
```

So the LEDscape server has a few options available. If you'd like to change the port used, change the `-p` value. `-c` stands for the number of pixels in each pin- change this to reflect the number of LEDs you are using. You're also going to add the `--no-demo` flag: this will turn off the pretty rainbows, but it will also

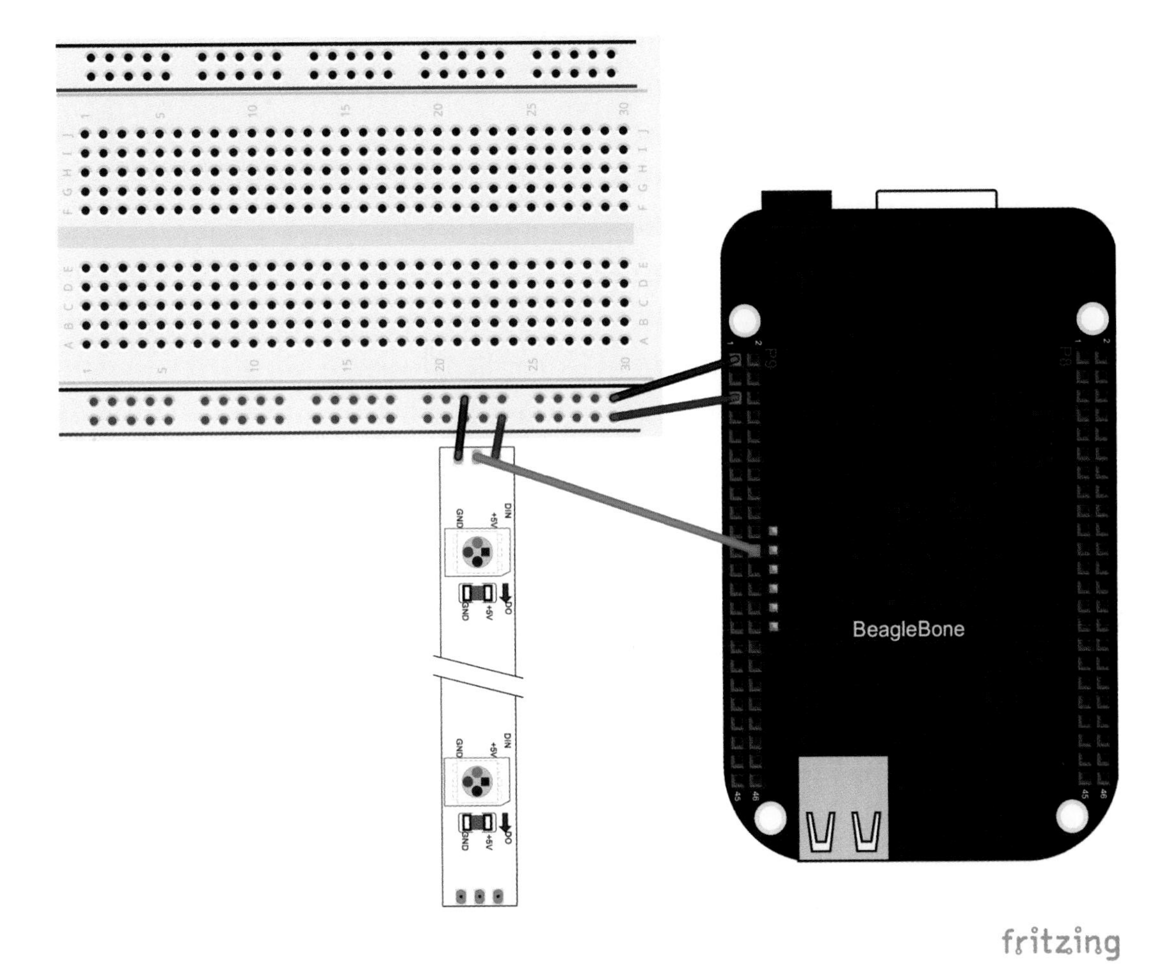

Figure 10-2 *Wiring diagram for the RGB LEDs to the BeagleBone Black*

allow you to stop sending signal from the Node programs you'll see shortly without those rainbows taking over. So, you're going to change the previous setup to a strip with 120 pixels, and you want to use port 7890:

```
./opc-server -p 7890 -c 120 --no-demo
```

Once you're done, save this file and exit your text editor (Control-O→Enter→Control-X if you're using nano).

5. Install the LEDscape service to run on boot.

   Now you'll set up the service. The commands shown in Example 10-4 apply to all models of BeagleBone Black.

Example 10-4 *Enabling the service*

```
sudo systemctl enable path/to/
LEDscape/ledscape.service
sudo systemctl enable ledscape
```

You'll need to change the path/to/ in your commands; for Rev C this should be /root/LEDscape/, for older models it should be /home/root/LEDscape/.

Nothing's happening! That's OK; you turned the demo off, so the server is still running, but just not sending data to the pixels. A quick way to check is to run the following in your *LEDscape/* directory:

```
./rgb-test
```

This script show rainbows again. It is a test script included with the LEDscape library.

Powering Many LEDs

So, if you're here, you're starting out with lots of LEDs. That's great! But powering them can be an issue. Definitely read up on Adafruit's power guide (*http://bit.ly/1bQOelW*) before continuing.

If you're using RGB-123 panels, use the power supply that the manufacturer sells when working with multiple panels—but this *does* require *advanced* soldering and electronics knowledge.

Code Time! Let's Bring in the JavaScript

All source code for the examples in this book can be found on GitHub (*http://bit.ly/19LX9n3*):

1. Installing dependencies:

   Node is already installed, luckily, on your BeagleBone Black. You do need to install a module globally so you can use it later. This module is called forever (*http://bit.ly/1bQOkdg*), and it allows you to run Node scripts in the background as services. To install, simply run the following:

```
npm install -g forever
```

If you get an error installing `forever`*, try prefixing the npm command with* `sudo`*, or try installing npm-sudo-fix (http://bit.ly/19LYYjT) first.*

2. Next, run the following commands on your BeagleBone Black:

```
cd ~/
mkdir my-led-project
cd my-led-project
npm install johnny-five beaglebone-io
open-pixel-control
touch open-pixel-test.js
```

This creates your project directory, changes into it, and uses npm to install three modules:

Johnny-Five
: A software layer that allows you to run JavaScript to control many different robotics platforms, with wrappers for each platform.

beaglebone-io
: The wrapper for BeagleBone Black—this will allow you to use all the friendly features of Johnny-Five with your BeagleBone Black.

open-pixel-control
: A module used to communicate with LEDscape using JavaScript.

Running a Test Script

Now that all of the software is installed, you can start writing some JavaScript.

There are two ways to handle editing your code: use your favorite terminal editor and edit directly on the BeagleBone Black, or use git, scp, or other tools to copy files between your computer and your BeagleBone. Try using git, but do what works for you.

Open up *open-pixel-test.js* and add the code shown in Example 10-5.

This connects your Node app to the LEDscape server, and once it's connected, it constructs a strip object. Open-Pixel-Control handles the math of multiple strips, so if you have multiple, go ahead and construct a strip for each pin (currently you need to to do this in order, but plans are in the works to improve this).

Then you construct an array of three-value arrays: the first value stands for red, the second is for green, and the third is for blue. Each array of three values represents the state of one LED on the strip. We'll talk about how to replace one pixel at a time in a moment.

Example 10-5 *open-pixel-test.js*

```
var opc_client = require('open-pixel-control');
var client = new opc_client({
  address: "127.0.0.1",
  port: 7890
});

//when the client is connected...
client.on("connected", function(){
  //create a model of the RBG LED strip we just
  //hooked up to the BeagleBone Black
  var strip = client.add_strip({
    length: 25
  });

  //create an array of pixel data (all soft white) to send to LEDscape
  var pixels = [];
  for(var i = 0; i < strip.length; i++){
      pixels.push([50, 50, 50]);
  }

  //This is the command to send the color data to our RGB LEDs
  client.put_pixels(strip.id, pixels);
});
```

Go ahead and exit your editor, copy your files to the BeagleBone if need be, and run:

```
node open-pixel-test.js
```

You should see a string of dim white lights!

That's cool and all, but wouldn't it be great if you could animate this? Definitely. So let's add an animation using the `put_pixel` method included with open-pixel-control. Add some code right after the `put_pixels` call in your *open-pixel-test.js* file, so it looks like Example 10-6.

What this does is, on an interval, lights up the next pixel in the chain to a random color. Go ahead and rerun the code:

```
node open-pixel-test.js
```

The strip should start with all dim white lights, but with each tick a light should spring up in a random color!

Example 10-6 *Add this code*

```
var opc_client = require("open-pixel-control");
var client = new opc_client({
  address: "127.0.0.1",
  port: 7890
});

client.on("connected", function() {
  var strip = client.add_strip({
    length: 25
  });

  var pixels = [];
  for (var i = 0; i < strip.length; i++) {
    pixels.push([50, 50, 50]);
  }

  client.put_pixels(strip.id, pixels);

  var index = 0,
    randomColor;

  //this time we're setting an interval to animate the new pixels
  setInterval(function() {
    //creating a random color to assign to a pixel.
    randomColor = [Math.floor(Math.random() * 256),
                   Math.floor(Math.random() * 256),
                   Math.floor(Math.random() * 256)];

    //this time, we're modifying one pixel at a time-the library
    //keeps track of the rest.
    client.put_pixel(strip.id, index, randomColor);

    index++;
    if (index == strip.length) {
      index = 0;
    }
  }, 1000);
});
```

Adding in Johnny-Five/ Beaglebone-io

So this is great! You have animation, you have lights, now to add some interaction. To do this, you're going to use the Johnny-Five API via the beaglebone-io wrapper. You'll also add a button that resets the animation.

First, wire a button to the BeagleBone Black. Then, modify the Johnny-Five code to interact with open-pixel-control. Go ahead and create a new file called *leds-with-button.js*, and copy your code from *open-pixel-test.js* into it so you can add to it.

Grab a button and wire it up: one side to ground, and one side to pin P9_39, as shown in Figure 10-3.

Now, you're going to add the beaglebone-io wrapper to the code, and pull in Johnny-Five to handle the button press, as shown in Example 10-7.

Example 10-7 *Add the beaglebone-io wrapper*

```
var five = require("johnny-five");
var BeagleBone = require("beaglebone-io");
var board = new five.Board({
  io: new BeagleBone()
});

var opc_client = require("open-pixel-con
trol");
var client = new opc_client({
  address: "127.0.0.1",
  port: 7890
});

//When the GPIO are ready to use
board.on("ready", function() {
  //construct a model of the button we just
wired to our BeagleBone Black
  var button = new five.But
ton("P9_39"); //A0 in Arduino Mapping

  client.on("connected", function() {
    var strip = client.add_strip({
      length: 25
    });

    var pixels = [],
      animationInterval;

    //initializes- clears the strip and
starts our random color animation
    reset_strip();
    start_animation();

    //when the button is pressed down
    button.on("down", function() {
      //clear the strip again and rerun the
animation
      reset_strip();
      start_animation();
    })

    function reset_strip() {
      clearInterval(animationInterval);

      pixels = [];
      for (var i = 0; i < strip.length; i+
+) {
        pixels.push([0, 0, 0]);
      }

      client.put_pixels(strip.id, pixels);
    }

    function start_animation() {
      var index = 0;
      animationInterval = setInterval(func
tion() {
        randomColor = [Math.floor(Math.ran
dom() * 256),

Math.floor(Math.random() * 256),

Math.floor(Math.random() * 256)];

        client.put_pixel(strip.id, index,
randomColor);

        index++;
        if (index == strip.length) {
          index = 0;
        }
      }, 100);
    }

  });
});
```

Just like other Johnny-Five applications, you have a `board.on('ready')` call. You construct a button, and call `button.on('down')` to wait for an actual button press, which resets the animation. I pulled out the reset and animation code into their own function so you don't have to duplicate code.

Go ahead and run it with this command:

```
node leds-with-button.js
```

Hopefully, your strip will start animating, then reset and begin again when you press the button.

Next, you're going to add some sensors and get some fun color combinations and brightness values using the environment around us!

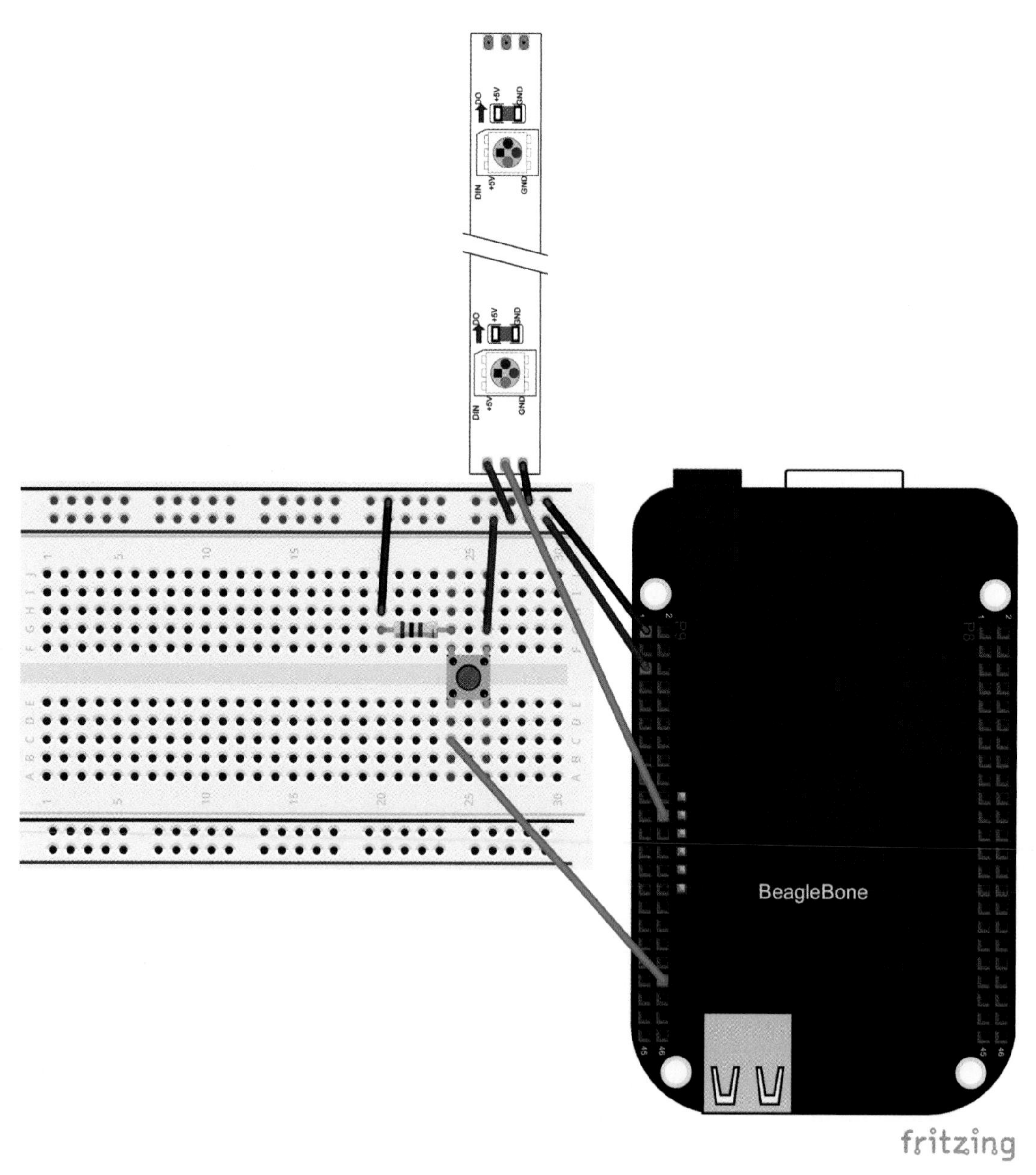

Figure 10-3 *Wiring diagram for the RGB LEDs and the button*

> **beaglebone-io Notes**
>
> Because BeagleBone Black has so many more GPIO pins than an Arduino, which Johnny-Five was originally designed for, you need a way to address the extra GPIO. The author of the library has given us two ways to address pins: one is a map from Arduino pin values to BeagleBone Black GPIO pins, or you can just mention the pins directly. In the sample code, the pins are addressed directly, but includes the Arduino-style mapping in comments when available. Use whichever is comfortable for you!

Adding a Photoresistor

First, let's use a photoresistor to determine how bright to make the LEDs. This is great for projects that want to run only in the dark, or want to scale the lights up when it's bright so they can still be seen. Go ahead and create a *lights-with-photoresistor.js* file for this in your project directory.

You can remove the button from the previous project if you wish—you'll wire the photoresistor to a new pin so you can leave the button in place if you want. Wire one end of the photoresistor to 5V, and one end to a 1k ohm resistor. On the row with the photoresistor and resistor, place your signal wire. On the row where your resistor ends, place your ground wire. Wire ground and 5V accordingly, and wire signal to P9_40. Figure 10-4 shows the wiring diagram.

Next, add the light data code shown in Example 10-8. In this version, we're going to scale the LEDs intensity up with the ambient lights—this makes the LEDs easier to see in the daytime.

So when the photoresistor receives data, the code sets the max value of the RGB values. Then, it displays white at that intensity to each of the pixels on the light strip. We use Johnny-Five's `scale()` function to map the photoresistor's output to a number between 0 and 255, which saves you some math.

Go ahead and test it out; run it just as you've done for all the other examples.

Changing Colors with an Accelerometer

Wire up your accelerometer along with your photoresistor—you're going to keep the light scaling code and sensor from before. A breakout board accelerometer is the easiest way to do this. I put mine on a separate breadboard so I can put extra long wires on it—I like to really wave it around to test it. I labeled the X, Y, and Z axis as purple, orange, and yellow, and they go to the P9_37, P9_38, and P9_35 pins, respectively. Figure 10-5 shows the wiring diagram.

So now that you can scale the brightness, let's have some fun with color. One of my favorite setup experiments is attaching a three-axis accelerometer to the inputs, and changing the value of R, G, and B with each axis. But first we need to think about accelerometer math. If your accelerometer tracks both negative and positive acceleration (i.e., both directions separately), we'll need to account for this. My accelerometer outputs from 0 to 1. When it's not moving, it's at .5—anything less is negative movement, anything more is positive.

You don't really care what direction it's going in—you just want faster acceleration to cause brighter color. So you're going to do some basic arithmetic, coupled with the `scale()` function from Johnny-Five. You'll map the input of each axis to a number from –1 to 1, then take the absolute value. This will map correctly to movement in either direction as a positive integer you can use, multiplied against the light values, to create a proportional color.

Then it's just a matter of reading each axis, mapping it to a color, and sending it out a strip (see Example 10-9). Want a challenge? Map each new color to a pixel, and have the results scroll by!

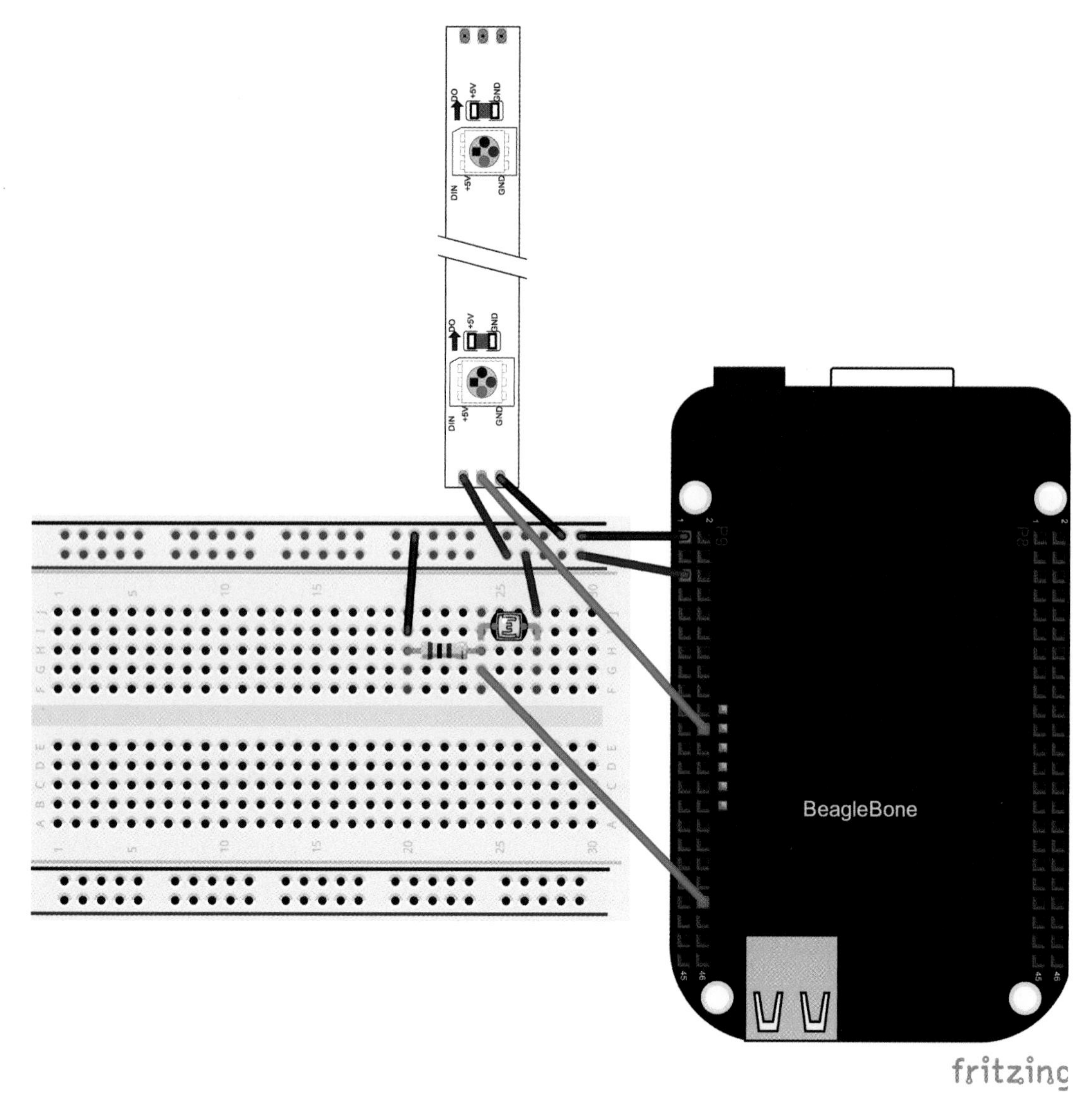

Figure 10-4 *Wiring diagram for the photocell and RGB LEDs*

What's Next?

There are so many things you can do with this new knowledge. Consider a *pixel photo booth*. For this, a Tessel (*https://tessel.io*) took pictures of people, a Node server scaled it down, and then sent the data to a BeagleBone Black, which used code similar to the code we wrote here to show the picture on a grid as shown in Figure 10-6.

Example 10-8 *Adding light data code*

```
var five = require("johnny-five");
var BeagleBone = require("beaglebone-io");
var board = new five.Board({
  io: new BeagleBone()
});

var opc_client = require("open-pixel-control");
var client = new opc_client({
  address: "127.0.0.1",
  port: 7890
});

board.on("ready", function () {
  this.digitalWrite(5, this.HIGH);
  var light = new five.Sensor("P9_40"); // A1 in Arduino Mapping

  this.repl.inject({
    light: light
  });

  client.on("connected", function(){
    var strip = client.add_strip({
      length: 25
    });

    var pixels = [];

    light.scale([0, 255]).on("data", function(){
      pixels = [];
      for(var i = 0; i < 120; i++){
        pixels.push([this.value, this.value, this.value]);
      }
      client.put_pixels(strip.id, pixels);
    });
  });
});
```

Example 10-9 *Mapping the input*

```
var five = require("johnny-five");
var BeagleBone = require("beaglebone-io");
var board = new five.Board({
  io: new BeagleBone()
});

var opc_client = require("open-pixel-control");
var client = new opc_client({
  address: "127.0.0.1",
  port: 7890
});

board.on("ready", function () {
  this.digitalWrite(5, this.HIGH);
  var light = new five.Sensor("P9_40"); //A1 in Arduino Mapping
  var accelerometer = new five.Accelerometer(["P9_37", "P9_38", "P9_35"]);
  this.repl.inject({
    light: light
  });

  client.on("connected", function(){
    var strip = client.add_strip({
      length: 25
    });

    var pixels = [];
    var maxValue = 0;

    light.scale([0, 255]).on("data", function(){
      maxValue = this.value;
    });

    accelerometer.scale([-1, 1]).on("data", function(){
      pixels = [];

      var xValue = Math.abs(this.x.value),
          yValue = Math.abs(this.y.value),
          xValue = Math.abs(this.z.value),
          red = xValue * maxValue,
          green = yValue * maxValue,
          blue = zValue * maxValue;

      for(var i = 0; i < strip.length; i++){
        pixels.push([red, green, blue]);
      }

      client.put_pixels(strip.id, pixels)
    });
  });
});
```

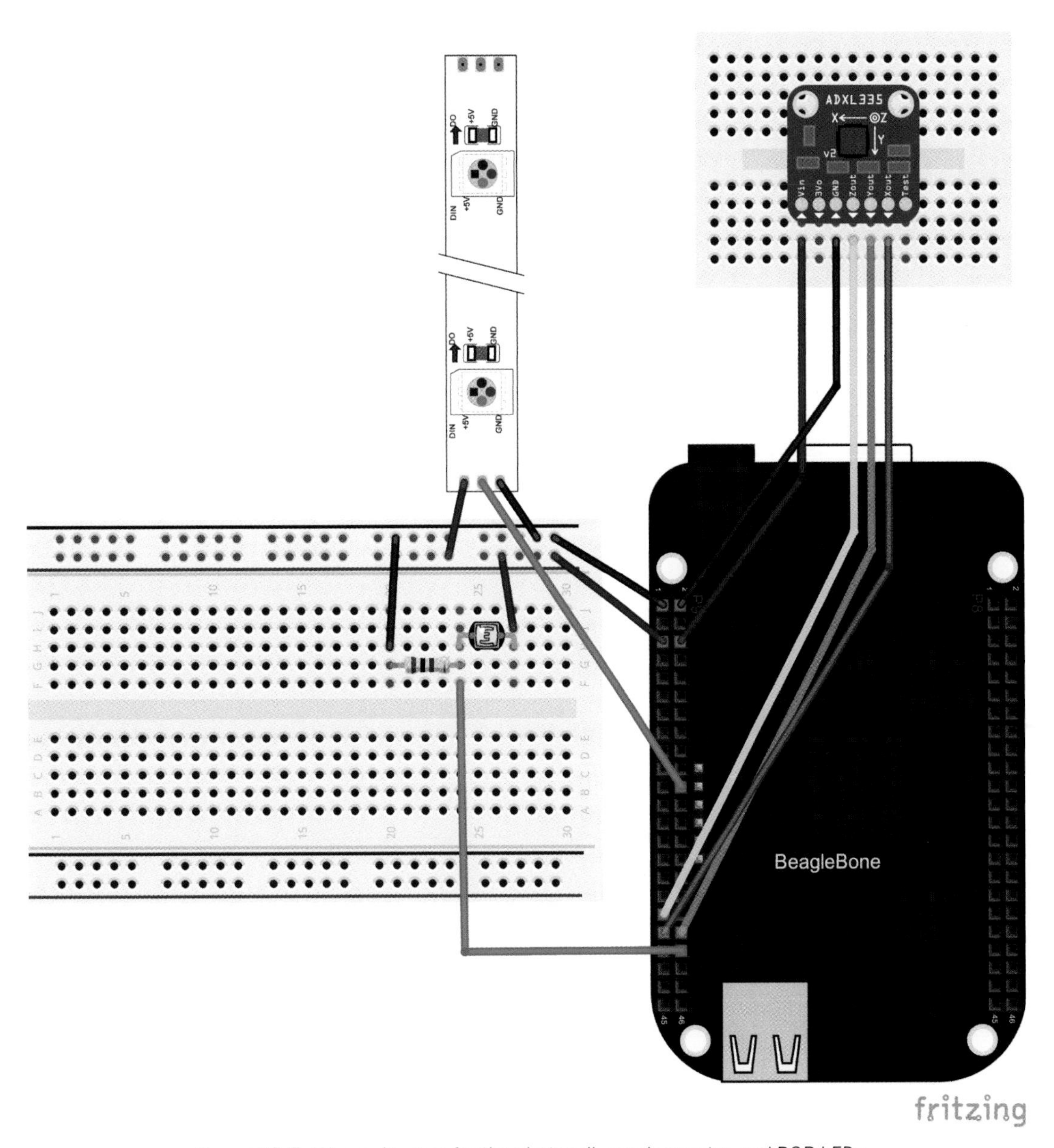

Figure 10-5 *Wiring diagram for the photocell, accelerometer, and RGB LEDs*

Figure 10-6 *Light grid showing the NodeBots logo*

There's also the entire Internet available. You can have your lights react to Twitter, Facebook, the weather, a software project's build status, email, or anything that has an API! Figuring out new ways to show data with color is a ton of fun with RGB LEDs.

Physical Security, JavaScript, and You

11

By Emily Rose

DIY physical security hacking can be challenging, yet also a joy to experience. This chapter will employ an approach of *progressive enhancement* to the development of a working security prototype that uses the following hardware:

- Arduino
- Laser diode
- 10k Ohm resistor
- Photovoltaic sensor
- HC-SR04 ultrasonic sensor
- 951WG magnetic contact switch

This chapter will be devoid of what you might consider an actual robot. The interesting aspect of implementing physical security systems is that although they may lack innate wow factor or a feeling of futurism, your creations can easily become a natural part of your daily routine.

One example would involve replacing the button of a garage door remote with a transistor wired to an Arduino. A Node.js application can then allow you to send SMS messages to the garage's dedicated phone number, which would toggle between opening and closing the door. Although completely mundane in its application, this is an example that is useful and easy to accomplish.

I cover a very similar use case in this chapter, but with more of a security focus than one of convenience (or laziness), as shown in Figure 11-1.

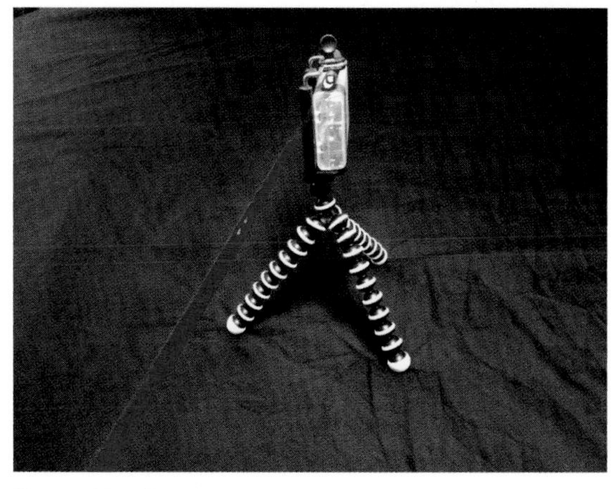

Figure 11-1 *Dramatic representation of completed project with battery-powered laser on tripod*

Simple Ultrasonic Sensor Project: Experimental Control Test (SUSPECT)

Let's start with the most basic circuit involving only the ultrasonic sensor known as the HC-SR04. The most notable thing about this particular sensor is its low price point and high availability. I typically pick these up by the handful for around $7 each on Amazon. They are also very flexible in their applications. They lend themselves to motion-detection gadgets, invisible measuring tapes, or even a DIY back-up sensor on that old Honda of yours! It never hurts to have a few of these lying around.

On to the relevant bits…

Implementation

Start by wiring the power (VCC) and ground wires directly to the respective pins for the sensor itself, as shown in Figure 11-2. Next, you will connect a wire from pin 11 on the Arduino, to either of the middle two pins on the *HC-SR04*. After carefully deciding which of the HC-SR04's center pins you want to wire to the Arduino, you may now now bridge the two center pins with an additional jumper. It is depicted in the diagram as a very small pink wire, but in reality you will most likely be using a much longer wire. This is perfectly fine for the purpose of prototyping!

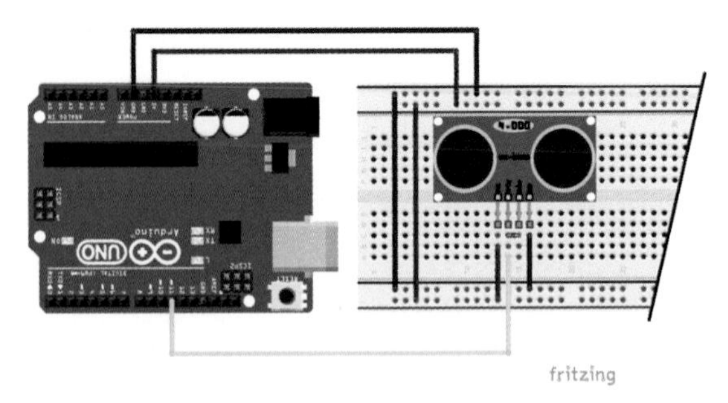

Figure 11-2 *Wiring the HC-SR04 and providing power to the rails of the breadboard*

> **Vigilance of Voltage Variance**
>
> Make certain you connect the 5V power pin on the Arduino to the power rails of the breadboard, and subsequently to the VCC pin on the sensor. Using the 3V3 power pin will result in unpredictable behavior, and it will be very confusing.

Once you have the schematic complete to the specifications of the diagram, you may turn your attention to the *small matter of programming* required of you prior to this new gadget coming to life. First, fire up the text editor of your choice, and pull up the nearest terminal emulator window. Upon creating a new directory for your lovely new gadget-to-be, you must now initialize the project. To do this, invoke the power of NPM via:

```
npm init
```

This command will immediately launch you into the most difficult problem known to computer science: naming a project. After intense deliberation, select a suitable name for your project, and continue. You don't need to worry about entering anything in particular on the remaining prompts (only the name matters for now); you're just doing this for the *package.json* file that is created.

Now that you have generated a *package.json* file for the little device, you must create your first file! Back to the text editor you go, so you may begin to craft the brains of our ultrasonic beast.

You must first `require` Johnny-Five, declare a `board` variable, and the variable for the ultrasonic sensor. The rest of the declarations are defining the variables you will use to track state for the tiny almost-robot and its sole sensor. See Example 11-1.

Example 11-1 *Declarations*

```
var five = require("johnny-five");
var init = false; // done taking baseline readings?
var trips = 0; // number of times the alarm has been triggered
var ultraSensor; // ultrasonic sensor to be represented by Ping object
var ultraBaseline; // baseline distance to measure triggers against
var ultraReadings = [ ]; // array of readings from ultrasonic sensor
var ultraThreshold = 4; // tolerance in inches before triggering alarm
var ultraTriggered = false; // state of ultrasonic triggeredness
```

Troubleshooting the Threshold

You may find that the initial `ultraThreshold` value results in many false-positives. A general rule is that the farther away the nearest object, the larger the required threshold. For distances up to 10′ or more, a threshold of up to approximately 12″ may be necessary for reliable alerts. Tweak this value until you find one that works for your application.

All source code for the examples in this book can be found on GitHub (*https://github.com/rwaldron/javascript-robotics*).

With initialization out of the way, you can now focus your efforts on the interesting parts, starting with this:

```
board = new five.Board(); ❶
board.on("ready", function() { ❷
  ultraSensor = new five.Ping(11); ❸
  ultraSensor.on("change", ultraChange); ❹
  ultraSensor.on("data", ultraData); ❺
});
```

We're only doing five things in the preceding code:

❶ Initializing our `board` variable as a new Johnny-Five `Board` object.

❷ Assigning the `ready` function as the listener, and for once the board is... *ready*.

❸ Once ready, initializing the `ultraSensor` variable as a new Johnny-Five `Ping` object.

❹ Assigning the `ultraChange` function as the listener to each `change` event from the sensor.

❺ Assigning the `ultraData` function as the listener to each `data` event from the sensor.

Poignant Point on Pin Particulars

Note the number passed as the only argument when we initialize our ultrasonic sensor `ultraSensor` as a Johnny-Five `Ping` object. It must correspond to the pin number labeled on the actual Arduino!

Everything is now "firing" so to speak, so all that's left is digging into the remaining three functions necessary to make everything really come to life:

`ultraChange`
: Handler for each change event from the `Ping` object, shown in Example 11-2.

`ultraData`
: Handler for each data event from the `Ping` object, shown in Example 11-3.

`trigger`
: Called when the sensor decides a significant event has occurred Example 11-4.

Example 11-2 *ultraChange() source code*

```
function ultraChange() {
  // Not initialized, do nothing
  if (!init) { ❶
    return;
  }
  var inches = this.inches;

  if (Math.abs(inches - ultraBaseline) > ultraThreshold) { ❷
    // if we haven't already triggered the alarm, do it!
    if (!ultraTriggered) { ❸
      trigger("ultrasonic"); ❹
      return ultraTriggered = true;
    }
  }
  ultraTriggered = false; ❺
}
```

Here we do several things:

❶ Check if we have yet been initialized; if so, bail immediately.

❷ Barring that, compare the current measurement to the base measurement with respect to the threshold we've set.

❸ If the value is beyond that of the threshold, then check if we've already triggered the alarm.

❹ If not; call the trigger function and return, setting triggered to `true` on the way out.

❺ If we don't determine the value falls outside of acceptable limits, ensure that `ultraTriggered` is set to `false` and go on our way.

Example 11-3 *ultraData() source code*

```
function ultraData() {
  var inches = this.inches; ❶

  if (ultraReadings.length >= 10) { ❷
    ultraReadings.shift(); ❸

    if (!init) {
      ultraBaseline = ultraReadings.sort()[4]; ❹
      console.log("Calculated baseline: %s", ultraBaseline);
    }
    init = true; ❺
  }
  ultraReadings.push(inches); ❻
}
```

Ah yes, our frequent-flyer function. Remember, this one is called (by default) once every 225 milliseconds. Despite the two layers of logic, this function is fairly straightforward:

1. First, declare the `inches` variable for convenience.
2. Next, determine if the number of readings sampled equals or exceeds 10.
3. If we do have more than 10 values in our readings array, shift one off the top.
4. If the `init` variable is `false`, we have hit our target number of readings from which to derive the baseline measurement. We do this by sorting the values, and picking one close to the middle.
5. Set the `init` variable to `true` (as long as you have at least 10 readings).
6. Finally, push the new value onto the end of the readings array and call it a day.

Essential Event Explanation

An important distinction to make here is the difference between the `change` and `data` events. The `change` event is only called when our instance of the `Ping` object is able to detect a difference between the current reading and the last. The `data` event is triggered continuously at a predetermined rate (which can be defined, and defaults to 225 ms), regardless of any change in value.

Example 11-4 *trigger() source code*

```
function trigger(sensor) {
  console.log("* Alarm has been triggered (%s) [%s]", sensor, trips);
  ++trips;
}
```

This brings us to the apex of our program, which is rather anticlimactic I'm afraid. This is where I leave it up to you: use your imagination! With the current implementation we simply increase the counter for the number of times the alarm has been triggered and log it to the console. Not much of an alarm really, now is it? Perhaps we can make it more interesting in iterations to come?

SMS Augmented Ultrasonic Sensor Application: General Experimentation (SAUSAGE)

So, you've successfully programmed the logic behind a very rudimentary ultrasonic sensor! Take a moment to congratulate yourself before you venture any further. There are a few potential problems with the design as it stands, but it's a good start!

Let's now focus on turning this little proof-of-concept into… you know, an actual alarm. As I mentioned previously, console logging does not an alarm make. You may be asking yourself how you might possibly build any usable alarm without some sort of loudspeaker or other such device. To this, I respond: you already (probably) carry one with you every day—a mobile phone! All you need is a little API magic from our friends at Twilio (*https://www.twilio.com*), and you have yourself a brand new alarm system! The best part of this is that it's completely free to debug and test with the Twilio API (although I certainly have found it quite useful—and inexpensive—to maintain an account balance on their platform).

I'll now give you some time to create a Twilio account if you haven't already done so. Go ahead, I'll wait...

Great! Welcome back! What's next? Well, you need to get this project set up with a Twilio API client library. Luckily for you, Twilio themselves have provided a pretty top-notch client just for happy Node users like us... Yay.

Implementation

So without further ado, I give you the ultrasonic SMS alarm! Let's get started: install the `twilio` library by navigating back to your project directory in your terminal emulator, and typing the following into the command line:

```
npm install --save twilio
```

That's it, you've just installed a Twilio client into your application! To verify this, you can check the contents of the *package.json* file to confirm that indeed the `twilio` library has been listed as a dependency.

And with that, you're off to the text editor to make some exciting new changes to the existing codebase! Let's begin with requiring `twilio` alongside `Johnny-Five`:

```
var five = require("johnny-five");
var twilio = require("twilio");
```

Along with requiring the library itself, you of course must do a little bit of configuration in order to make use of this wonderful API. Fortunately for you, the only configuration you need consists of two strings, which are made available to you from the Twilio dashboard once logged in (Figure 11-3). Simply copy and paste from the dashboard into your code like so:

```
var client = twilio(YOUR_TWILIO_SID,
                    YOUR_TWILIO_AUTH_TOKEN);
var lastSMS = 0;
var ratelimit = 5000;
```

Figure 11-3 *Locating your Twilio account SID and auth token from the user dashboard*

With these additions, you're almost ready to start sending SMS messages by waving your hands in front of some electronics! You need only add a few more lines of code before you launch your very first SMS! Let's revisit the cute little trigger function and give it some bite. Change the trigger function as shown in Example 11-5.

Example 11-5 *The modified trigger() function*

```
function trigger(sensor) {
  var now = Date.now();

  console.log("* Alarm has been triggered
(%s) [%s]", sensor, trips);

  if (now - ratelimit < lastSMS) {
    return console.log("> Ratelimiting.");
  }

  ++trips;
  lastSMS = now;

  client.messages.create({
    body: "Alarm has been triggered by "
     + sensor,
    to: "YOUR RECIPIENT'S PHONE NUMBER",
    from: "YOUR SENDER'S PHONE NUMBER"
  }, function smsResults(err, msg) {
    if (err) {
      console.log("*** ERROR ***\n", err);
      return;
    }

    if (!msg.errorCode) {
      console.log("> Success!");
    } else {
      console.log("> Problem: %s",
       msg.errorCode);
    }
  });

  console.log("> Sending SMS.");
}
```

There are a few things to explain in the preceding code. It instructs the Twilio client to create a new message, and passes it an object containing a few important values:

`body`
: The text body of the message you'd like to send

`to`
: The SMS-capable phone number that will receive the message

`from`
: The number that will be shown on caller ID

The Twilio Sandbox and Phone Number Formatting Follies

Twilio is particular with number formatting, so make sure you use the exact format shown. Also worth mentioning is that the `from` number used here is a Twilio sandbox number that is free for testing. You can find the best sandbox number to use for your area code in the Twilio user dashboard. Just make sure you either verify your own number or buy a Twilio number before going into production.

Now you find yourself in control of a coarsely tuned alarm system that's actually capable of alerting you to potential intruders! All without any circuit modifications from the preceding SUSPECT exercise. From here on out, it just gets crazier, so now may be the time to play with what you already have and get comfortable with the current set of moving parts. Seems sufficiently simple so far, right?

Point-of-Entry Monitoring System (PoEMS)

> The magnetic switch,
>
> a tried-and-true solution.
>
> Security now!

At this point, you have a working implementation of an actual alarm system. Let's make it a bit more reliable and robust by way of adding one of the most prominent components of any real security system: the magnetic contact switch. This overwhelmingly simple mechanism has been a staple of security systems for eons. It is of course not without its limitations, but if you're looking to monitor events such as window breaches or door traffic, you can't get much easier than this. The anatomy of the magnetic switch is simple. You are presented with a pair of wires protruding from an enclosure (metal or plastic). This enclosure is joined by a mate that usually appears similarly, sans wires (or possibly just a magnet). You may not be able to tell, but bringing these two parts into contact causes the connection between the two wires to be closed. Pulling the partner piece away from the wired piece opens the connection, and thus triggers an alarm. This basic mechanism is what makes it so reliable. There are no microprocessors involved in its operation. Aces.

Implementation

Adding this component presents you with your first opportunity to modify the circuit you previously completed in the SUSPECT exercise. Fear not, as this is an extremely simple addition to the existing work (Figure 11-4). You merely add two jumper wires and the magnetic switch to the breadboard, and you're done with hardware for this round. You don't even need to worry about polarity with these switches!

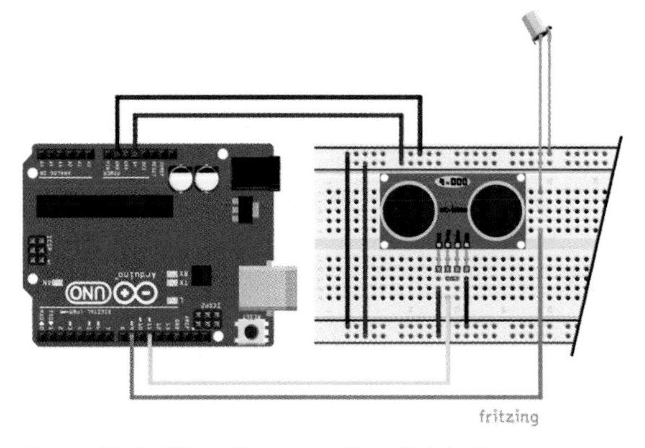

Figure 11-4 *Wiring the magnetic switch to the breadboard*

As for additional programming, you only need to declare two new variables, and register *lis-*

teners for state changes. Logically, this switch performs identically to a button, and Johnny-Five provides you with the ability to declare "pullup" buttons, which is perfect for what you need to do with the switch:

```
var magnetSensor;
var magnetTriggered = false;

board.on("ready", function() {
  ultraSensor = new five.Ping(11);
  ultraSensor.on("change", ultraChange);
  ultraSensor.on("data", ultraData);

  magnetSensor = new five.Button({ ❶
    isPullup: true,
    pin: 12
  });
  magnetSensor.on("up", function() { ❷
    trigger("magnet");
  });
});
```

Here are the changes:

❶ Add initialization of the `magnetSensor` variable as a new instance of the `Button` object, making sure to pass the `isPullup` property as `true`.

❷ Begin listening for the *up* event (which in this case is triggered when the partner magnet is physically separated from the wired portion). You could also listen for the *down* event here, but you're only interested in knowing when the connection has been broken.

Simple code changes;

New button and event, done!

This is Johnny-Five.

It's almost hard to believe, but that's it! You've successfully built a magnetic-contact-switch trigger into your alarm system. The project is getting more useful with each successive iteration, but still it feels as though something is missing, don't you think? Let's unlock this further…

Lasers Impress Both Enemies and Relatives, Thank You (LIBERTY)

The title of this section is your complimentary canned response to the inevitable questioning you will receive after building this project:

"What's with the lasers all over your house?"

Don't get me wrong; you owe nobody any explanation for lasers. *Lasers are their own explanation.* Not only are they incredibly impressive, they're also relatively simple to deploy and can be very noticeable when placed in conspicuous locations and the correct environment. They are also the perfect excuse to break out the fog machine! Security theater never looked so good.

Implementation

To create a laser trigger, you need only three things:

- 5mw 650nm laser diode
- Photovoltaic sensor
- 10k Ohm resistor

An optional (but extremely helpful) component to this project is some sort of *sleeve* to place the sensor into. This helps cut down on false positives from changes in ambient lighting. I have found a piece of heat shrink tubing to work well for prototyping. The only other materials required for this project are those related to how you decide to mount the alarm. Take a trip to your local hardware store and let your imagination run wild.

Let's take a look at the modifications to our diagram. We are again faced with very limited additions to the circuit. We're adding only the laser diode, and the wiring necessary to read analog values from the photovoltaic sensor, as shown in Figure 11-5.

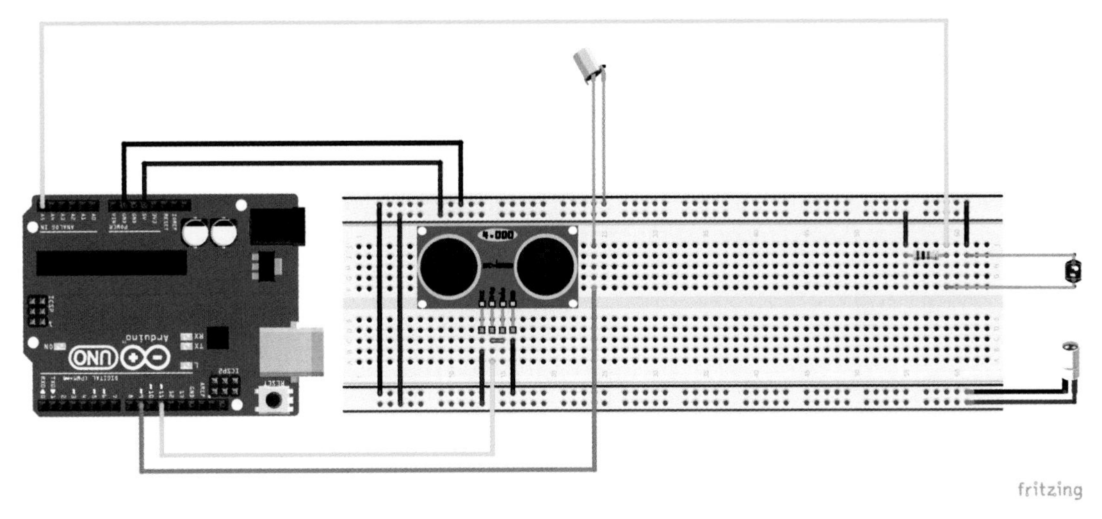

Figure 11-5 *Wiring the laser diode, photovoltaic sensor, and 10k Ohm resistor to the breadboard*

An interesting thing to note is that we're not actually controlling the voltage to the laser module. This means that you don't need to worry about what pin it's on or write any code for it! Assuming you're using a laser diode that can accept 5V, there's no reason not to wire it directly to the power rails, or even power it with a completely separate power supply!

When working with the photovoltaic sensor it's important to be mindful that although the laser is much brighter than typical ambient lighting, other sources of light can still cause false alarms. To mitigate this possibility, place the sensor inside a sleeve of some sort. You can use heat shrink tubing in a pinch (and for prototyping). This creates for the sensor a rudimentary *cone of acceptance* that will significantly reduce the opportunity for interfering light. The laser diode can also be given the same treatment for more discreet laser-based monitoring applications.

As for code modifications, you may already be ahead of me at this point but let's lay it all out just to be safe. Adding to the ever-growing header of variable declarations, Example 11-6 now includes the various variables necessary for your functioning photovoltaic sensor, powered by laser!

Example 11-6 *The laser-ified header*

```
var ultraSensor;
var ultraBaseline;
var ultraReadings = [ ];
var ultraThreshold = 4;
var ultraTriggered = false;

var magnetSensor;
var magnetTriggered = false;

var photoSensor; // photovoltaic sensor to be represented by Sensor object
var photoReading = 0; // most recent reading from the photovoltaic sensor
var photoThreshold = 100; // difference between light readings to tolerate
var photoTriggered = false; // state of photovoltaic triggeredness
```

```
var client = twilio(YOUR_TWILIO_SID, YOUR_TWILIO_AUTH_TOKEN);
var lastSMS = 0;
var ratelimit = 5000;
```

Along with the additional variable declarations, you of course must add to the setup magic in the `ready` function, and create a function named `photoData` to handle incoming values from the sensor, as shown in Example 11-7.

Example 11-7 *New functions*

```
board = new five.Board();
board.on("ready", function ready() {
  ultraSensor = new five.Ping(11);
  ultraSensor.on("change", ultraChange);
  ultraSensor.on("data", ultraData);

  magnetSensor = new five.Button({
    isPullup: true,
    pin: 12
  });
  magnetSensor.on("up", function() {
    trigger("magnet");
  });

  photoSensor = new five.Sensor("A0");
  photoSensor.on("data", photoData);
});

function photoData() {
  var data = this.value;
  if (Math.abs(data - photoReading) > photoThreshold) {
    if (!photoTriggered) {
      trigger("laser");
    }
    photoTriggered = true;
    return;
  }
  photoTriggered = false;
  photoReading = data;
}
```

Your new `photoData` function should look very familiar after having written the `ultraData` and `ultraChange` functions. With the photovoltaic cell being a simple analog sensor, a simplified algorithm will more than suffice. You're employing the concept of a threshold here as well, and again calling `trigger` if a new reading falls outside of the acceptable variance.

Status Indicator Necessary, Buttons and Diodes (SINBaD)

The defensive functionality of our dear device is just about complete, but we are still noticeably lacking in the user interface department. What can we do to remedy this situation? Buttons and LEDs, that's what! In all reality, you will most likely be more interested in a secure

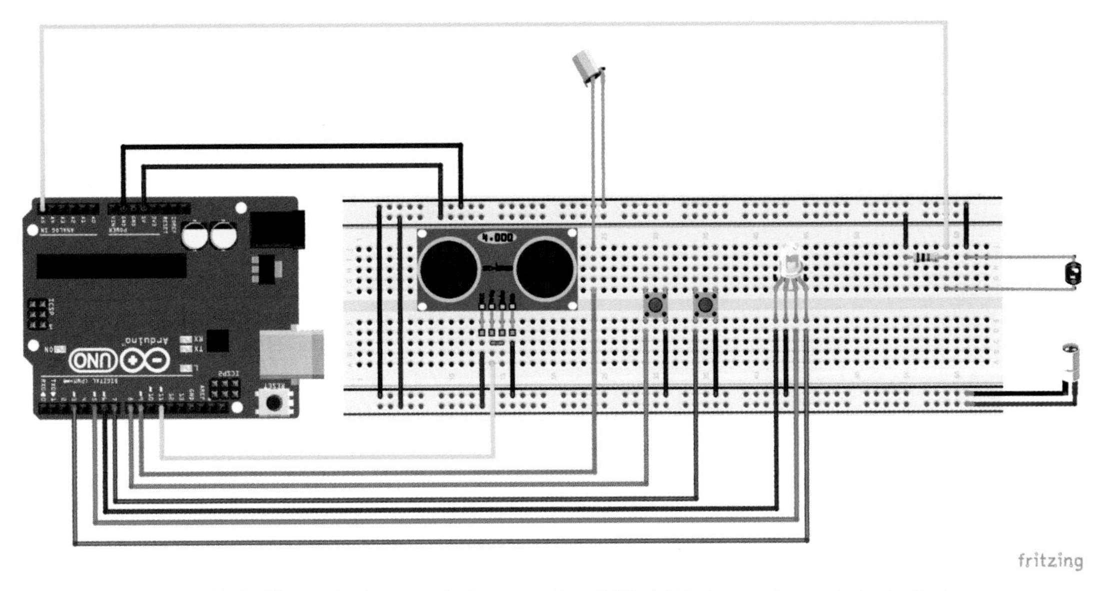

Figure 11-6 *Wiring the two new buttons, and an RGB LED to be used as a status indicator*

> **Pulse Problems Prevent Progress**
>
> You may notice that we are skipping pin 4. This is done in case you later decide you want to make the LED fade/pulse, or emit custom hues of light. This requires pulse-width modulation (PWM), which pin 4 does not support.

mechanism to arm and disarm your system, but I'll leave that to you to sort out. Secure systems should avoid allowing the user to directly alter system state without some sort of authentication.

Let's completely disregard that for the sake of simplicity by implementing an indicator LED, a reset button for the ultrasonic sensor, and an arm/disarm button! You may want to explore adding remote arm/disarm functionality via the Twilio API when you're ready to take it to the next level (just make sure you only accept input from authorized phone numbers).

Implementation

The two new momentary push buttons will behave just like the existing magnetic contact switch, but in reverse! Whereas the magnetic switch is closed (current flowing) until the magnets are separated; the momentary switch buttons are open until depressed. We will again be opting for pullup buttons, so as to avoid the hassle of reading more of those cryptic resistor bands! In this case, we'll wire the two buttons up to pins 7 and 8. The RGB LED we've added is of the kind known as "common cathode"; each of the three colors of LED have their own positive lead, and share the same ground lead. Figure 11-6 shows how to connect the leads for red, green, and blue to pins 6, 5, and 3 respectively.

Obviously these new buttons also represent some additional code. You will again be implementing Johnny-Five pullup `Button` objects for the new, actual buttons—finally! The RGB LED is also covered with a special superset of the `Led` object. As always, first start with declaring our new variables up top, as shown in Example 11-8.

Example 11-8 *New variables*

```
var armButton; // armDisarm button to be represented by Button object
var resetButton; // resetUltra button to be represented by Button object
var status; // status indicator LED to be represented by Led.RGB object
var isArmed = false; // prevent alarm from triggering when device is disarmed
```

As for the actual functionality of our new components, make the following modifications to the `ready` function, as well as introducting the `ultraReset` and `armDisarm` functions. Your code should look something like Example 11-9 after you're done.

Example 11-9 *The modified code*

```
board = new five.Board()
board.on("ready", function ready() {

  armButton = new five.Button({
    isPullup: true,
    pin: 8
  });
  armButton.on("up", armDisarm);

  resetButton = new five.Button({
    isPullup: true,
    pin: 10
  });
  resetButton.on("up", ultraReset);

  status = new five.Led.RGB({
    pins: {
      red: 6,
      green: 5,
      blue: 3
    }
  });

  ultraSensor = new five.Ping(11);
  ultraSensor.on("change", ultraChange);
  ultraSensor.on("data", ultraData);

  magnetSensor = new five.Button({
    isPullup: true,
    pin: 12
  });
  magnetSensor.on("up", function() {
    trigger("magnet");
  });

  photoSensor = new five.Sensor("A0");
  photoSensor.on("data", photoData);

})
```

```
function ultraReset() {

  console.log("* Resetting...");
  ultraReadings = [];
  init = false;
}

function armDisarm(override) {

  if (typeof override == "boolean") {
    isArmed = override;
  } else {
    isArmed = !isArmed;
  }

  if (isArmed) {
    console.log("* Arming");
    status.color("#FF0000");
  } else {
    console.log("* Disarming");
    status.color("#00FF00");
  }
}
```

Now that you have completed your additions to the `ready` function, and added the functions associated with the two new buttons; you may also modify two other functions to make use of the new user interface. To start, add the following as the very first line of the `trigger` function to prevent the alarm from triggering when the device has been disarmed:

```
if (!isArmed) { return; }
```

It's also useful to indicate via LED when the ultrasonic sensor is gathering data for the `ultraBaseline` value. If this sounds like it will be useful for your purposes, you can modify your `ultraData` function, as shown in Example 11-10, to make the indicator LED blue when calibrating the ultrasonic sensor upon initialization, and when you press the reset button. We've also modified the function to automatically rearm the device once the ultrasonic sensor has finished recalibrating by calling `armDisarm(true)`, which is optional.

Example 11-10 *Modified ultraData()*

```
function ultraData() {
  var inches = this.inches;
  if (ultraReadings.length >= 10) {
    ultraReadings.shift();

    if (!init) {
      ultraBaseline = ultraReadings.sort()[4];
      console.log("Calculated baseline: %s", ultraBaseline);
      armDisarm(true);
    }
    init = true;
  } else {
    statusLight.color("#0000FF");
```

```
    }
    ultraReadings.push(inches);
  }
```

What's Next?

At this point, you are now in possession of a physical security prototype that contains a somewhat complete suite of functionality. From here, you can of course make your own customizations and improvements to dramatically increase the practical usefulness of your new device. The first recommendation is to obtain and house your project within one of the project boxes depicted in Figure 11-7. You may also explore the option of converting your project from breadboard to protoboard as you begin to finalize the design and scope of your particular implementation.

Figure 11-7 *The project box that houses Arduino, the breadboard, and applicable wires*

The possibilities are quite endless, and the material covered here has only just begun to scratch the surface (Figure 11-8 shows our finished product).

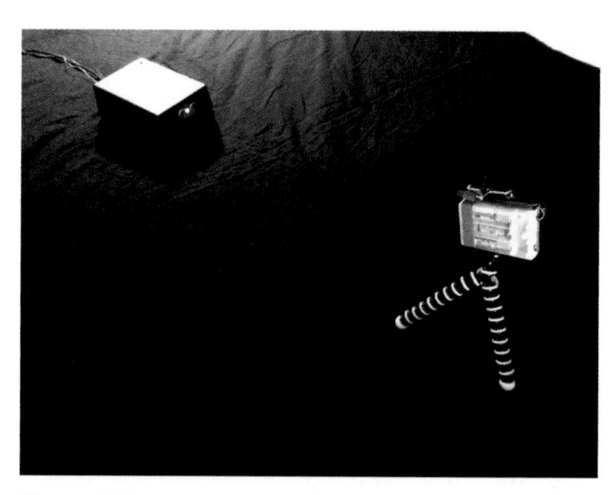

Figure 11-8 *Our final security prototype in action*

For those of you who have taken interest in this chapter, feel free to use this project as a springboard for new and more powerful ideas. I am extremely interested in feedback, and look forward to hearing about your creations. Potential collaborators are welcome to engage with me on this subject at the GitHub repository (*http://bit.ly/1zEHFHa*) for this project.

12 Artificial Intelligence: BatBot

By Raquel Vélez

Have you ever wondered if robots will take over the world? If so, you're not alone. Hollywood and the media have done an excellent job of trying to convince us of an impending robot revolution.

To complete such a mission, robots would need to be smart enough to work together and develop a plan to take over the world. To accomplish this, however, they would need some human-enabled intelligence, also known as artificial intelligence. Fortunately, we are a long way off from enabling robots to work together at such a large (and certainly scary!) scale, but artificial intelligence is still very useful for all sorts of applications, and that is why we're going to spend this chapter playing with it!

Figure 12-1 *BatBot*

In this chapter, I'm going to talk about artificial intelligence (AI) and three categories of artificially intelligent robots (remote-controlled, semi-autonomous, and fully autonomous). And then you're going to build BatBot, as shown in Figure 12-1, a semi-autonomous robot.

Artificial Intelligence: The Basics

As humans, our ability to make decisions and act on them is what has propelled us to the top of the intellectual food chain. Our abilities to reason, rationalize, and remember give us the leg up on other species who can't do it quite as well.

When we talk about artificial intelligence, all we really mean is using algorithms to simulate decision making. An artificially intelligent being (i.e., a robot) relies entirely on these algorithms to know what to do in certain situations. In most circumstances, robots don't learn like humans do; they have to be taught absolutely everything.

Machine learning is the one exception to the AI rule. To be clear, however, even machine learning has its limitations. Machine learning is simply a series of increasingly complex algorithms (like Haar Cascades in neural networks) that can be used to "teach" a robot how to discern different pieces of its environment, to which it can apply other decision-making models.

I like to group robots into one of three categories, ranging from least to most artificially intelligent: remote-controlled, semi-autonomous, and autonomous.

Remote-Controlled Robots

Remote-controlled (RC) robots have no artificial intelligence. They are completely controlled by a human, and thus their decision making is non-existent. Humans make every decision for the robot.

Examples of remote-controlled robots include RC cars, simple circuits, and computer numerical control (CNC) machines. There is a direct correlation between what the human wants to do and what the robot does, for better or for worse (if a human makes a mistake, the robot makes a mistake, too!).

Semi-Autonomous Robots

Semi-autonomous robots have some artificial intelligence. They get most of their instructions from a human, but also make some of their own decisions based on factors in their environment about which the human either doesn't know or doesn't care to know.

Examples of semi-autonomous robots include the Mars rovers, da Vinci surgical robots, and Parrot AR drones. Humans give the robots the important instructions ("go explore that area"), but the robot has enough sensors and artificial intelligence to make its own decisions within those parameters (like avoiding obstacles or dampening subtle movements due to shaking or wind turbulence).

Autonomous Robots

An autonomous robot is controlled entirely by artificial intelligence. After receiving some initial instructions from a human, it makes all of its own decisions. Many people like to argue that there are no fully autonomous robots in existence yet, as most robots still require a significant amount of human intervention. However, subsystems of semi-autonomous robots can be fully autonomous.

Examples of semi-autonomous robots with autonomous components include the Google self-driving car, the Nest thermostat, and commercial airplanes (think autopilot). After being given a mission ("drive to X" or "keep my house at Y degrees"), they use their sensors and artificial intelligence to figure out how to complete their task.

BatBot

To really play with artificial intelligence, we're going to teach BatBot how to find its way out of a paper bag. It's a silly challenge, yes, but the lessons that result from this exercise are the foundations for any artificially intelligent robot.

BatBot is a semi-autonomous robot: we'll remotely control it (so that we can put it in the bag), and then tell it to find its own way out using its ultrasonic (sonar) sensor. As hinted by its name, BatBot will use sonar to navigate its way out of the paper bag. To add a bit of fun, we will use a Playstation DualShock Controller to navigate it remotely via Bluetooth, and use XBee radios to keep the robot separate from the computer.

Bill of Materials

Table 12-1 *Bill of materials for the sonar sensor array*

Count	Part	Source	Estimated price
1	MaxBotics Ultrasonic Rangefinder LV-EZ2	AF 980; SF SEN-08503	$25
1	Generic high-torque standard servo	SF ROB-11965; AF 155	$12
1	100 ohm resistor	Electronics retailers	$10 for a variety pack
1	100 uF capacitor	Electronics retailers	$1.50
	Headers	Online Electronics retailer	$2 for a variety pack
	Jumper wires	MS MKSEEED3; SF PRT-11026; AF 758	$2 for a variety pack
1	XBee wireless kit	SF KIT-13197	$100
	Glue gun and glue sticks	Online retailer	$10

Table 12-2 *Bill of materials for the chassis*

Count	Item	Source	Estimated price
1	Arduino Uno	MS MKSP99; AF 50; SF DEV-11021	$25
1	BOE Bot Robotics Shield Kit for Arduino Uno	MS MKPX20; SF ROB-11494	$130
5	AA batteries	Online retailer	$5
1	PS3 or PS4 DualShock controller	Online retailer	$45

Table 12-3 *Additional (optional) materials*

Item	Source	Estimated price
Bat wings, cut out of felt or similar	Local craft store	Varies

Item	Source	Estimated price
A ridiculously large paper bag, like a lawn paper bag for disposing of autumn leaves, with most of the sides cut down	Your local hardware store	$3 for a pack of five
Decorative accessories (e.g., googly eyes, a feather boa, etc.)	Local craft store	Varies

Some Notes About the Materials

The focus of this chapter is making robots smarter, with more of an emphasis on software than on hardware. As such, while I assume you have the materials listed, you also have completely free reign on the materials you choose to use for your robot. For example, instead of using an Arduino Uno, you can use a SainSmart Uno R3. Instead of a BOE Bot, you can use a SimpleBot (described in Chapter 1) or a SumoBot Jr. (*http://sumobotkit.com*)

You may also skip the XBee wireless kit and simply use a very, very long USB cable (the longest you can find!), and replace the large paper bag with three sturdy walls made of non-signal-absorbent material (Styrofoam is especially bad for this project). If you choose to use a different type of ultrasonic sensor, refer to the Johnny-Five-sanctioned list of sonar sensors (*https://github.com/rwaldron/johnny-five/wiki/Sonar*). Simply put, feel free to get creative—the really interesting bits are when you get to the software portion!

Many of the smaller components like the resistor and the capacitor can be acquired with some of the other components: for example, the BOE Bot Robotics Kit includes such components, as do many Arduino starter kits.

There will be some soldering required for this project. If you don't feel comfortable with your soldering skills, you should practice a bit or ask for assistance. The soldering isn't particularly complex, but it does require a bit of finesse.

Assembly

Let's start building!

1. First, assemble your chassis according to the manufacturer's instructions. We won't be making any special modifications to the chassis except adding to it, so building it should be fairly straightforward.
2. Solder some headers to the sensor (Figure 12-2) , so that you can easily fiddle with your connections. The most important connections for this project are GND (ground), +5 (voltage in), and AN (analog output signal).

Figure 12-2 *Ultrasonic sensor with headers before soldering*

3. Attach your sonar to a servo horn using hot glue (Figure 12-3): it's easy to take

apart if you mess up, but holds really well. As an added bonus, it won't damage any of your components, and your bot shouldn't be heating up hot enough at any point to risk the glue melting again.

Figure 12-3 *Ultrasonic sensor with servo horn, attached with hot glue*

4. Attach your standard servo to the front of your bot using hot glue, as shown in Figure 12-4. Be sure to attach the servo in such a way as to ensure that the sonar will rotate left to right, pointing ahead of the robot.

Figure 12-4 *Servo attached to batbot*

5. Wire up the sensor array according to Figures 12-5 and 12-6, which will also require the 100 ohm resistor and the 100 uF capacitor.

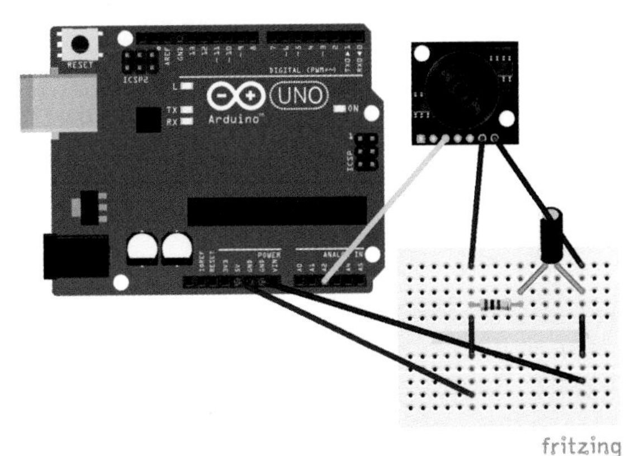

Figure 12-5 *Fritzing diagram of sonar sensor array*

Dirty Power versus Clean Power

Because you're using several electronic components (in particular, the XBees and the ultrasonic rangefinder), there will be electromagnetic interference. This interference occurs because the XBees are emitting electromagnetic radiation (so they can talk to each other). As a result, the ultrasonic sensor's readings are degraded.

The technical term of this degraded performance due to electromagnetic interference is called "dirty power."

To clean our power up, we regulate the voltage coming into the sonar by using a resistor and a capacitor. When you have a dirty power situation, an insufficient amount of voltage is coming through to the sensor. The capacitor "cleans" the power by sitting in the line of voltage, filling up with whatever voltage comes through (charging), and releasing the right amount of voltage (discharging) to the sensor. For those who like analogies, the capacitor works like a stock room at a grocery store —shipments for produce will come in at different times, but the shelves will always be well stocked. Thus, a shopper (the sensor) won't need to know (or care) if eggs are being delivered today, as long as there are some available on the shelves.

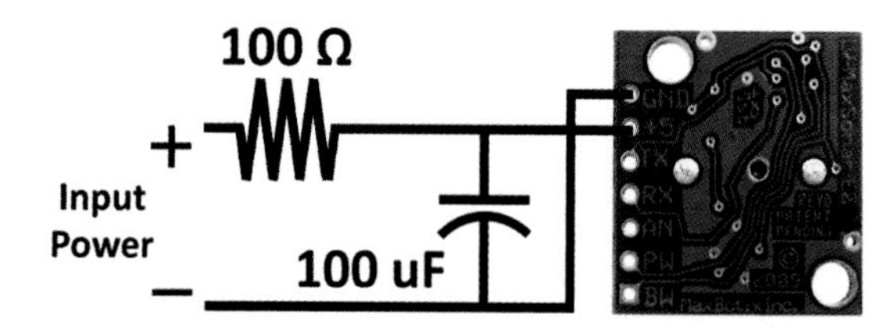

Figure 12-6 *Schematic diagram of sonar sensor array*

6. Once you've attached the sonar, now is a great time to add some finishing touches. Add a pair of wings, flame stickers, or whatever suits your fancy. Just make sure it doesn't get in the way of the wheels or the sonar; you want to make sure the robot can still make readings and move freely so it can finish its task!

Now that you have all of the major bits in place, let's get to the meat of this project: artificial intelligence!

Step 1: Remote Control

Before you can get to the really awesome fun part of making BatBot find its way out of a paper bag, you're going to need to figure out how to talk to BatBot:

1. Ensure you have the latest stable version of Node.js and npm installed on your computer. If you still need help with installing Node.js and need a primer on how to use npm to install modules, see the appendix.
2. Get the code for BatBot, located in the *batbot/* folder in the *Make: JavaScript Robots* repository on GitHub (*http://bit.ly/1N65wXs*).
3. On your local copy of the code, find your way into the *batbot/* directory and run `npm install` to install all of the packages listed in the *package.json* file. I'll introduce each module as you need them. The first and most important one is `johnny-five`, which allows you to send commands to the Arduino and thus move the servos and read from the sensor.

Moving the Robot

Now that you have a robot and your software environment is set up, the next major task is to get BatBot moving around under your direction. From there, you can move on to encouraging BatBot to drive itself.

Let's take a closer look at the chassis.

Notice that the BOE Bot comes with two continuous servos, one for each wheel. Each wheel moves independently, which will allow the robot to move in any direction: forward, backward, left, and right.

*A **continuous servo** moves continuously in a single direction (i.e., clockwise), like a motor. Where it differs from a standard motor, however, is that we can programmatically tell it to move in the opposite direction (i.e., counterclockwise). (To make a standard motor switch directions, on the other hand, you would have to physically change the polarity of its inputs.)*

As you can see in Figure 12-7, the two servos are pointed in opposite directions. This means that in order for the robot to move in a straight line, the servos are going to have to turn in opposite directions (i.e., one will turn clockwise while the other turns counterclockwise). Keep in mind, though, that both wheels will still turn in the same direction.

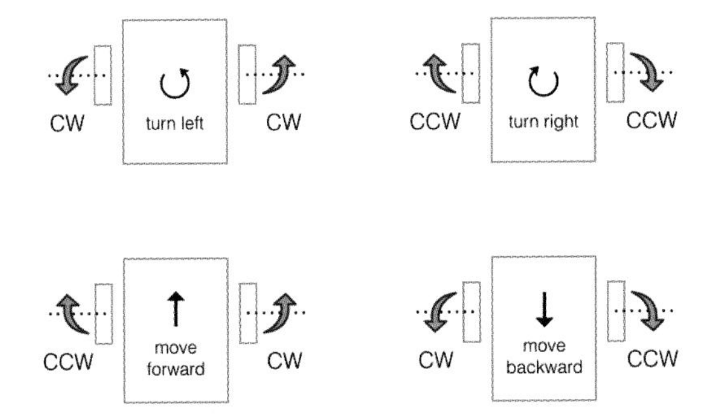

Figure 12-7 *Simplified diagram of robot movement*

Similarly, if both servos turn in the same direction, the robot will turn! For this project, when the robot turns, you want it to turn in place. To do this, one wheel needs to turn backward at the same rate that the other wheel turns forward.

By implementing each servo separately, you have the logic shown in Table 12-4 for moving the robot.

Table 12-4 *Continuous servo logic for robot movement*

Direction of Movement	Left Servo Direction	Right Servo Direction
Forward	Forward (ccw)	Forward (cw)
Backward	Backward (cw)	Backward (ccw)
Left	Backward (cw)	Forward (cw)
Right	Forward (ccw)	Backward (ccw)

Remember, all of the source code for the examples in this book can be found on GitHub (*http://bit.ly/19LX9n3*). You'll need to follow these steps:

1. Go ahead and create a new file in the *batbot/* directory called *moveBot.js*. Initialize `johnny-five` and begin our program like so:

```
var five = require("johnny-five");

var board = new five.Board();
board.on("ready", function () {
    // do stuff
});
```

2. Next, implement each continuous servo using the `johnny-five` servo API (*http://bit.ly/1bQONfn*). Add the following code after requiring the `johnny-five` module, but before initializing the board:

```
var leftServo =
   five.Servo.Continuous(10);
var rightServo =
   five.Servo.Continuous(11);
```

   The pin number corresponds to the connection of each servo to the BOE Bot shield and thus to the Arduino. You must also specify that these servos are continuous servos, as opposed to standard servos.

3. To make it easier to control each servo, write a `move` function, following the logic for robot movement described in Table 12-4:

```
var moveSpeed = 0.1;

function move(rightFwd, leftFwd) {
  if (rightFwd) {
    rightServo.cw(moveSpeed);
  } else {
    rightServo.ccw(moveSpeed);
  }

  if (leftFwd) {
    leftServo.ccw(moveSpeed);
  } else {
    leftServo.cw(moveSpeed);
  }
}
```

4. For a given movement, you want the right wheel to move forward or backward, and the same for the left wheel. Your code uses booleans to dictate the direction of each wheel. With this, you can abstract each movement out even

further with easier-to-remember functions:

```
function turnLeft() {
  move(false, true);
}

function turnRight() {
  move(true, false);
}

function goStraight() {
  move(true, true);
}

function goBack() {
  move(false, false);
}
```

5. Don't forget to include a `stop()` function as well:

```
function stop() {
  lServo.stop();
  rServo.stop();
}
```

6. You can play with the servos and move functions in the `johnny-five` REPL by passing them into the `johnny-five` REPL object:

```
this.repl.inject({
  left: leftServo,
  right: rightServo,
  turnLeft: turnLeft
});
```

and then in the `johnny-five` REPL:

```
leftServo.cw();

leftServo.stop();

turnLeft();
```

REPL

REPL is short for read-eval-print loop. It gives you the ability to input things at the command line, then have code evaluate that input, and finally output the results and then wait for the next command.

Controlling the Robot

Now that you have the wheels hooked up with `johnny-five`, let's add the PS3 DualShock Controller. The `dualshock-controller` module will allow you to map different keys on the controller to specific events.

1. Add the `dualshock-controller` module into the mix. Depending on your model of controller, you may need to change `dualShock3` to `dualShock4`:

```
var five = require("johnny-five");
var dualshock =
  require("dualshock-controller");

ds = dualshock({
  config: 'dualShock3'
});
```

2. You can make it so that when you press the triangle button, the robot moves forward in a straight line, via the `goStraight` function. Remember, this goes inside the `board.on("ready", ...)` block:

```
ds.on("triangle:press", function () {
  goStraight();
});

ds.on("triangle:release", function ()
{
  stop();
});
```

3. Map as many buttons to whatever functions suit your fancy. Remember to stop the robot when you let go of a button! Otherwise the robot will continue forever and ever. By only moving on keypresses, you will have much more control of your robot.

With your DualShock Controller mapped to your servos, you should now be able to remotely control your robot!

Try it out—how does it feel?

Pointing and Reading from the Sonar

Now let's hook up the sonar and its associated standard servo into the mix:

1. Because the sonar is an analog sensor, you need to wire it up to one of the analog pins on the Arduino:

   ```
   var sonar = new five.Sonar("A2");
   ```

2. Throw the sonar into the REPL, and start playing around with the readings, using `sonar.cm` or `sonar.inches`. What happens when you point it at different materials? Do you notice a pattern in readings? Is there a minimum reading or a maximum reading? You can read more about the `johnny-five` sonar API on GitHub (*http://bit.ly/1bQOYav*).

3. You should notice that the sensor emits higher values for objects that are farther away. You may also notice that the object doesn't necessarily need to be directly in front of the sonar to get a reading. This is due to the MaxBotix sensor's beam characteristics.

For maximum control of the sonar sensor, it is attached to a standard servo. Unlike a continuous servo, which moves continuously, a standard servo moves to a specified angle measurement. The benefit of a standard servo in this application is that we can specify exactly where we want the sonar to point. We can see the `johnny-five` servo API on GitHub (*http://bit.ly/1bQONfn*).

1. Initialize the sonar servo:

   ```
   var sonarServo = new five.Servo({
     pin: 12,
     range: [10, 170]
   });
   ```

   The `range` parameter allows you to set a minimum and maximum angle for the sonar servo; this way, instead of having to constantly remember which angle is "left," "center," and "right," you can simply say `sonarServo.max()`, `sonarServo.center()`, and `sonarServo.min()`, respectively.

Depending on how you've mounted your servo to the robot, `sonarServo.max()` *and* `sonarServo.min()` *may mean right and left, respectively. This is perfectly fine; just be sure to make the adjustments to your code as necessary.*

2. To finish, map the servo movement and sonar readings to your DualShock Controller:

   ```
   var angle = 15,
     sonarStep = 10;

   ds.on("r2:press", function() {
     console.log(sonar.cm);
   });
   ds.on("l2:press", function() {
     angle = (range[0] + range[1]) / 2;
     sonarServo.center();
   });
   ds.on("dPadLeft:press", function() {
     angle = angle < range[0]
       ? range[0] : angle + sonarStep;
     sonarServo.to(angle);
   });
   ds.on("dpadRight:press", function() {
     angle = angle > range[1]
       ? range[1] : angle - sonarStep;
     sonarServo.to(angle);
   });
   ds.on("dpadUp:press", function() {
     angle = range[1];
     sonarServo.max();
   });
   ds.on("dpadDown:press", function() {
     angle = range[0];
     sonarServo.min();
   });
   ```

   The r2 button press gives you sonar measurements, while the l2 button

Ultrasonic Sensor Quirks

To understand why the sonar is working the way that it is, take a look at the specification sheet (*http://bit.ly/1bQP02n*). Take extra care to understand what the sheet is saying about the sensor: the beam characteristics change between the different models, and will result in different readings.

Unlike a laser range finder, which has such a thin beam that it's virtually linear, a sonar uses a conical beam to gather data. The sonar's poor resolution makes it a cheap option for detecting walls, but not fine details. Knowing the needs of your project and the capabilities of different sensors is critical to choosing the right sensor for your project!

centers the servo. Then you're using the direction pad to incrementally move the servo in steps of `sonarStep` (and stopping at the minimum/maximum we set when we initialized it). You're also using the direction pad to move entirely to the maximum and minimum ranges. You're keeping track of the angle yourself to ensure you have the maximum amount of control.

Be careful when driving your bot around—make sure it's on a flat surface, preferably on the floor. If you must drive it around on a table, make sure you have someone standing guard who can catch the bot if (when!) something goes wrong.

Drive it around the room—how does it handle? Would you like to do anything differently? Feel free to play around, tweaking numbers. Make it your own!

Step 2: Autonomy

The next step on your journey to artificial intelligence is to take your remote-controlled robot and make it smart! To achieve this goal, you need to fully understand the problem at hand, break it down into smaller problems, "teach" the robot to handle those smaller problems, and walk away once the greater problem has been solved. The steps you follow now will be useful for any artificial intelligence problem, beyond helping BatBot find its way out of a paper bag.

Start by clearly identifying the problem: you have a paper bag, situated on the floor. The opening is pointed out, so that the robot can drive into the bag. Once in the paper bag, its task is to navigate its way back out, using only the ultrasonic sensor and its driving mechanism.

The robot goes into the bag, pointing at the back wall. How should it get out?

If you're having trouble seeing the world from the robot's perspective, pretend that *you* are the robot. Imagine that you are blindfolded, or the room is very dark. The only information you have is that there is or isn't a wall in front of you. On top of that, you can only turn 90° or move forward/backward. Now how do you get out of the room?

Your first thought may be to turn around by 180° and walk out.

While that's a perfectly valid answer, you're using *a priori* data (facts you knew before you walked into the room, like knowing that the room has three sides and you walked in through the open end). The robot doesn't have that information.

The goal of this exercise shouldn't be to answer this specific question, but instead to answer a general set of problems. This problem of the paper bag is essentially three walls, but it can be easily extended to solving a simple maze.

A common maze-solving algorithm is the wall follower algorithm (*http://bit.ly/1bQP6GY*), also known as the lefthand rule or the righthand rule. The general idea is that by following a wall with either your left or right hand along the wall, you will eventually find the end of the puzzle.

But going back to this problem's limitations, you only know if there's a wall in front of your eyes, not if you're parallel to a wall (though you can play around with that idea in a future iteration!).

It's important to note that the robot has no idea about where it is relative to anything else. It only knows the information it has at a particular moment, with no sense of memory.

For this specific application, then, there is an even simpler algorithm:

1. Check if there is a wall to the left of me, in front of me, and to the right of me.
2. If one of the directions has no wall, turn 90° or drive toward the opening; go back to step #1.
3. If there isn't, turn 90° to the right and go back to step 1.

Robot State

I'm touching on a concept called *state* without actually going into too much detail. The concept of state is important: a robot's *state* is the data the robot knows about itself at a given time. Robots rely on their sensors to gather this data, and then they can use it to get more information over time. Typical pieces of data that contribute to a robot's state are its position in space, heading, pitch, roll, velocity, acceleration, a map of its surroundings, a list of local landmarks, and more. BatBot has no understanding of any of these things; the only sensor it has is its sonar. Thus, for the autonomy portion of its semi-autonomous nature, it can only use its single sensor to understand its environment.

By using this pattern, you don't need to have any information about the room, and you can use the sensors you have available to make decisions.

Open Loop and Closed Loop

This specific algorithm is called an *open control loop*; it runs in a loop and doesn't ever correct itself. A more robust algorithm would be a *closed control loop*; it runs in a loop but uses the data it collects to adjust its next iteration, like a thermostat. I don't cover closed loops in this project, but no course in artificial intelligence is complete without it.

Don't worry if this algorithm isn't perfect—your first algorithm rarely, if ever, is. But at least you've got a place to start; from here, you'll push it to the robot and test it. Then, once you've identified the problems or limitations of the algorithm, you can tweak it and adjust it to your heart's content!

Implementing the Algorithm

Now that we've settled on an algorithm, let's write it up in the code:

1. First, check to see if there is a "wall" to the right of the robot. Point the standard servo to the right with `sonarServo.max()` and take a sonar measurement with `sonar.cm`.

   Next, turn the servo to the front (`sonarServo.center()`), take a measurement, and finally to the left (`sonarServo.min()`) with a measurement as well:

   ```
   sonarServo.max();
   var rightVal = sonar.cm;
   sonarServo.center();
   var frontVal = sonar.cm;
   sonarServo.min();
   var leftVal = sonar.cm;
   ```

2. Bind the start (and stop!) of the scanning algorithm to buttons on your DualShock Controller, to make it easy to put your robot in (and out of) autonomous mode:

```
ds.on('select:press', function () {
  console.log('IN AUTO MODE');

  var loop = setInterval(function () {
    ds.on('r1:press', function () {
        clearInterval(loop);
    });

    // your algorithm goes here
  });
});
```

Try it out and see what happens. What kind of values are you getting?

It's not quite working, is it? It turns out that, with this piece of code, the sonar readings are happening too fast for the servo to keep up.

JavaScript, as a language, is unique in that it is asynchronous by nature. When it sends a command, it doesn't wait for the command to complete before sending the next one. In the case of the servo/sonar combination, you're sending the commands in succession, almost instantaneously, and you're moving on to the next command before its predecessor has had a chance to complete.

What you want to do, instead, is give each command as much time as it needs to complete before beginning the next command. As a result, for this specific application, you're going to have to force the algorithm to be synchronous.

Fortunately, there's a library called `temporal` that will allow you to specify when each servo movement/sonar measurement takes place:

1. Add the `temporal` package to your code:

   ```
   var five = require("johnny-five");
   var dualshock =
     require("dualshock-controller");
   var temporal = require("temporal");
   ```

2. Using `temporal`, create a queue that moves the servo and takes a measurement every 1,500 milliseconds (i.e., 1.5 seconds):

   ```
   var scans = [];
   temporal.queue([
     {
       delay: 0,
       task: function () {
         sonarServo.max();
         scans.push({ dir: "left",
                      val: sonar.cm });
       }
     },
     {
       delay: 1500,
       task: function () {
         sonarServo.center();
         scans.push({ dir: "center",
                      val: sonar.cm });
       }
     },
     {
       delay: 1500,
       task: function () {
         sonarServo.min();
         scans.push({ dir: "right",
                      val: sonar.cm });
       }
     }
   ]);
   ```

   You may have noticed that now you're pushing our sonar measurements into an array. By using the `array-extended` module, you have some very useful utilities for manipulating arrays and extracting useful data.

3. Add the `array-extended` module to the code.

4. Take the array of three directional measurements and find the one that is mostly likely to be the open wall (given that higher sonar measurements indicate the wall is farther away):

   ```
   var maxVal = array.max(scans, "val");
   ```

5. Now use that information to implement the rest of the algorithm:

   ```
   WALL_THRESHOLD = 15; // cm

   var direction =
     maxVal.val > WALL_THRESHOLD
     ? maxVal.dir : "right";
   ```

```
if (direction === "center") {
  goStraight(1000);
} else if (direction === "left") {
  turnLeft(700);
} else {
  turnRight(700);
}
```

The WALL_THRESHOLD is the value at which anything below it implies that there is a wall present; anything above it is too far away and can be considered an opening.

6. To improve accuracy, take multiple scans in each direction and average them out using the array-extended module:

```
var scanSpot = function (cb) {
  var sServoReadings = [];
  var read = setInterval(function () {
    sServoReadings.push(sonar.cm);
    if (sServoReadings.length === 10)
    {
      clearInterval(read);
      cb(null,
        array.avg(sServoReadings));
    }
  }, 100);
};
```

Here's what's going on: when you call scanSpot(), you're taking a servo reading every 100 milliseconds (i.e., one-tenth of a second), and logging that in an array. After 10 sonar measurements, you use array-extended to find the average, and return that value via the callback. The callback ensures that you wait for this function to finish before you move on to the next step in the algorithm.

Put together, Example 12-1 shows the complete algorithm.

Example 12-1 *Finished algorithm*

```
var scans = [];
temporal.queue([
  {
    delay: 0,
    task: function () {
      sonarServo.max();
      scanSpot(function (err, val) {
        scans.push({ dir: "left", val: val });
      });
    }
  },
  {
    delay: 1500,
    task: function () {
      sonarServo.center();
      scanSpot(function (err, val) {
        scans.push({ dir: "center", val: val });
      });
    }
  },
  {
    delay: 1500,
    task: function () {
      sonarServo.min();
      scanSpot(function (err, val) {
        scans.push({ dir: "right", val: val });
      });
```

```
      }
    },
    {
      delay: 1500,
      task: function () {
        WALL_THRESHOLD = 15;
        minVal = array.min(scans, "val").val;
        var maxVal = array.max(scans, "val");
        var direction = maxVal.val > WALL_THRESHOLD ? maxVal.dir : "right";
        if (direction === "center") {
          goStraight(1000);
        } else if (direction === "left") {
          turnLeft(700);
        } else {
          turnRight(700);
        }
      }
    }
  ]);
```

You're going to want to repeat all of this indefinitely, or until you tell it to stop. Take a look at the *sonarscan.js* file for the complete version.

Time to try it out! Drive your robot into the paper bag and turn on autonomous mode! How does it do? Feel free to make adjustments until your robot achieves success.

Troubleshooting

My XBees aren't communicating—what's going on?
: Make sure your have them properly configured. See GitHub (*http://bit.ly/1C2ATMe*) for more information.

Why does the robot sometimes not listen to my DualShock Controller commands?
: First, make sure that you're intentionally stopping/starting the robot in your code. If you're absolutely positive that the code is good, it might be your XBee or Bluetooth connection. XBees are known to have some packet loss, but generally if you send another command it will set itself right again. Try replacing the XBees with a USB cable. If everything works perfectly, there's either something wrong with the XBee connection or your hardware setup. If that's not it, check your code again.

My robot isn't seeing the opening of the paper bag
: Check to make sure the top of the bag isn't dipping into the conical beam of the ultrasonic sensor.

What's Next?

Congratulations! Your little BatBot can now, on its own, find its way out of a paper bag, as shown in Figure 12-8!

Figure 12-8 *BatBot's movin' on out!*

While artificial intelligence requires quite a bit of concentrated thinking, it also really takes your robots to the next level. Want to go deeper? Try some of these exercises to go further in your artificial intelligence mastery:

- Instead of making the robot turn in place, make it turn in an arc
- Make the robot find its way out of a longer paper bag
- Make the robot solve a maze
- Implement the wall follower (*http://bit.ly/1bQP6GY*) algorithm
- Find and implement more interesting/complex algorithms to solve this puzzle
- Make a robot that avoids obstacles
- Add other sensors to the robot to make it even "smarter"

Delta Robots and Kinematics

13

By Pawel Szymczykowski

Robots can do a lot of neat things like beeping, rolling around, chasing your dog, or bringing you a cold beverage. That is one kind of robot —a "fun time" robot. The other kind of robot is hard working, precise, and industrious. These are the kinds of robots that welded your car together and perfectly placed the fake gem in the center of your treasure troll's belly. Welcome to the world of industrial robots! In this chapter, I'll examine one of the most common industrial

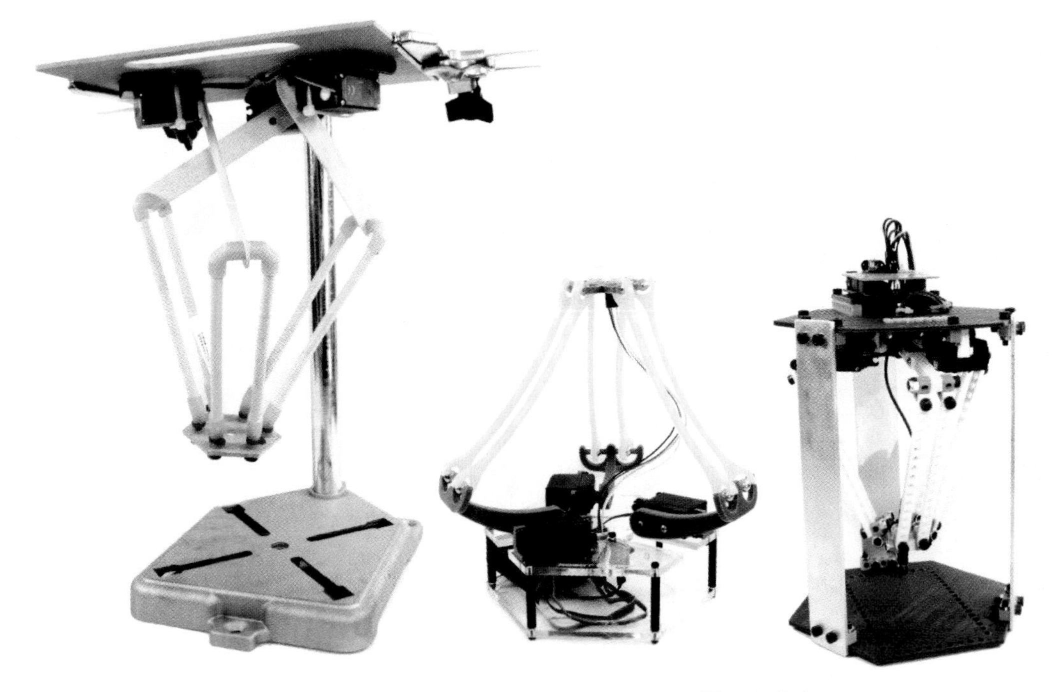

Figure 13-1 *Junky Delta, Robot Army, and TapsterBot*

robot designs, the delta robot shown in Figure 13-1.

A delta robot is a stationary robot that uses three multi-jointed arms connected to a central platform that it can move around in three dimensional space by shifting the positions of each of the three arms. That sounds like a mouthful! The truth is that delta robots are fairly simple mechanically. They use relatively few components and are easy to build compared to other types of Cartesian robots, which need special pulleys, linear bearings, lead screws, and more. On the flip side, they are harder to program and require some more complicated mathematics to control precisely. You probably guessed there would be some trigonometry involved when you saw the word *delta*, and you were right!

The delta was invented in 1985 by Reymond Clavel, a Swiss mechanical engineer, for the purpose of loading chocolate pralines into their packaging. Because the heavy motors are fixed at the top and because they only have to make small movements to move the arm a great distance, delta robots can move very quickly and accurately. They are well suited for packaging where they can quickly and accurately pluck items off of a moving conveyor belt, orient them, and place them into a package or assembly in fractions of a second.

Another common use is in pick-and-place machines that place tiny surface mount components onto circuit boards for manufacturing electronics. In recent years, a variant of the delta that uses lead screws to extend the reach of the *Z* axis has become a popular design in the hobbyist 3D printing industry.

In this chapter, I'll show you how to build a simple delta robot out of hardware store parts. If you happen to have access to a 3D printer, you might want to check at the end of the chapter for two alternative designs that you can 3D print to save some time and effort. Once you've built the bot, I'll show you how to wire it up and make the motors move individually. Finally, I'll get into some kinematics so you can position the arm precisely.

Bill of Materials

Most of the things listed in Table 13-1 can be found at a home improvement store. Locally, in the United States, you can try Home Depot or Lowe's, but any home improvement or hardware store will do!

You will need the items listed in Table 13-1.

Table 13-1 *Bill of materials*

Count	Item	Source	Estimated price
1	4ft length of 5/16" hardwood dowel	Hardware store	$1
1	2ft length of 1/4" inner diameter latex tubing or surgical tubing	Hardware store	$10
6	5/16" bolts	Hardware store	$0.25 each
6	#4 screws	Hardware store	$1
6	Small zip ties	Hardware store	$3
3	Large zip ties	Hardware store	$3

Count	Item	Source	Estimated price
1	8.5″ × 11″ piece of hardboard	Hardware store	$5
3	180° rotation standard servos (HS-311)	Amazon	$10 each
3	Jumbo-sized Popsicle sticks or paint stirrers	Amazon or hardware store	$2 or free
Arduino Uno or similar	Maker Shed, Amazon, Adafruit, Spark Fun	$30	1
4 AA battery case	Amazon	$1	Breadboard
Amazon	$3	Breakaway headers or jumper wires	Amazon
$5	9	Cardboard pieces cut into 1.5″ × 1.5″ squares	A dumpster

Here are the tools you need:

- Power drill
- 5/16″ drill bit
- 1/16″ drill bit
- Coping saw or scroll saw
- Spray adhesive or white glue
- Duct, gorilla, gaff tape
- Scissors

Delta Anatomy

Before you get started building, let's take a look at what parts make up a delta robot, shown in Figure 13-2. This will aid you in assembly, and help you understand how they work!

End effector
: The end effector is also sometimes called the tool platform. It's the business end of an industrial robot where you will find the gripper, cutter, vacuum print head, paint nozzle, or whatever the correct tool for the job is!

Universal joint
: A typical joint moves in two directions. A universal joint or u-joint can bend in any direction. You might be familiar with them if you've ever seen a drive shaft on a car. There are a number of alternative ways to create a universal joint, from bolting plate together, to friction fit ball bearings, to magnetic joints, and more.

Lower arm
: The lower arm connects the end effector to the upper arm. Each lower arm is actually made of up a pair of rods that each have their own joints, but they always move in parallel. This forms a parallelogram between the end effector and upper arm. It's a very important feature of the delta design, because without the parallelograms, the end effector couldn't stay parallel to the base and would skew in relation to its position. That would make the design much less useful.

Upper arm
: The upper arm is connected to the lower arm and the actuator. It's usually shorter than the

lower arm and only moved back and forth with the position of the actuator.

Actuator
: An actuator is a fancy engineering term for a motor that moves something. It can be any kind from an electric stepper motor to a hydraulic piston. The delta robots you will be looking at all use 180° rotation servo motors, which are inexpensive and easy to control.

Building Junky Delta

Junky Delta (shown in Figure 13-3) is the easy-to-build and inexpensive delta robot with a fun and zany self-deprecating name. I chose to optimize on ease of assembly over precision, so you won't need a 3D printer, CNC mill, or laser cutter to assemble him, but you also won't be moving individual atoms around. You might even have trouble moving around M&Ms here. Junky Delta can be assembled for around $50, mostly from parts you can get at a home improvement store. If you do have access to a 3D printer, check the end of this chapter for a couple of other options that might work better!

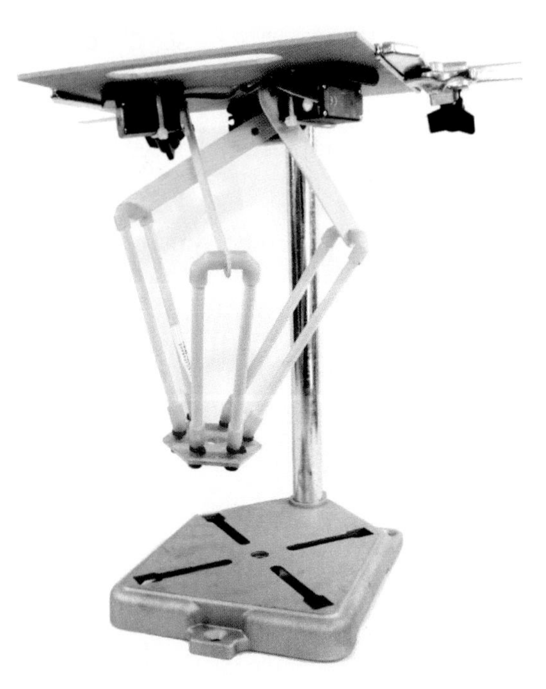

Figure 13-3 *Junky is serviceable and made of woody bits*

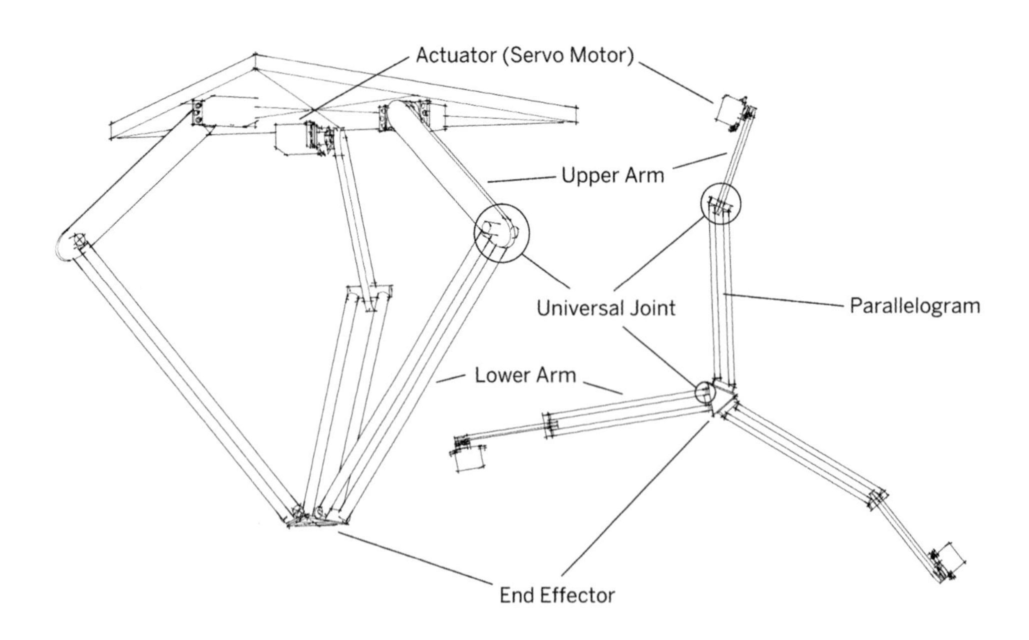

Figure 13-2 *Anatomical diagram of a delta robot*

You can download 3D printer files (*http://bit.ly/1bQPghE*).

1. First, cut the dowel into six 6″ long pieces and three 1″ long pieces using your saw or a pipe cutter, as shown in Figure 13-4.

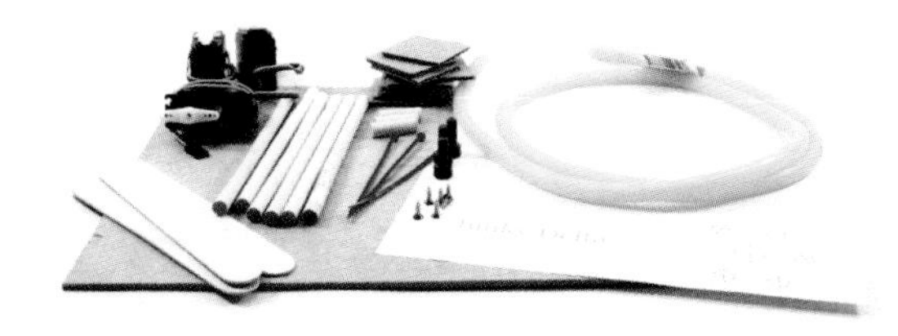

Figure 13-4 *Parts for Junky Delta with dowels pre-cut*

2. Next, you'll want to cut the latex tubing into twelve 1″ long pieces (Figure 13-5). These will hold your dowels together and form a universal joint. If you can't find latex tubing, something else will work as long as it's soft enough to bend easily and fits snugly over the dowel. You can size the dowels up or down as needed too, as long as they are thick enough to not bend easily.

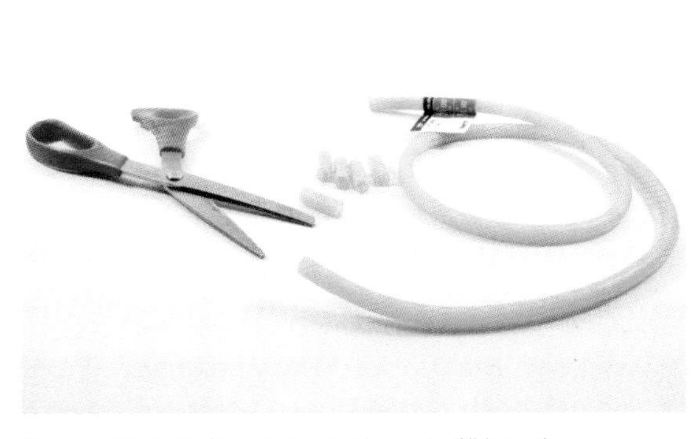

Figure 13-5 *Cutting latex tubing into 1″ lengths*

3. Get the template PDF file from the GitHub repository, print it out, and affix it to the hardboard. Now cut out the end effector and the extra bits from the board using your saw (see Figure 13-6). You don't have to be fancy and cut out the hexagon—you can make it square or circular (you can use a large hole saw) and it will work just as well.

Figure 13-6 *Cutting hardboard*

4. Using the template as a guide, drill out the holes indicated by the red circles with crosses in them using the 5/16″ bit, as shown in Figure 13-7. If you're as bad at drilling as me, you'll want to drill pilot holes in the center of the pluses using the smaller bit first.

Figure 13-7 *Drilling holes*

5. Now pop your bolts through the outer holes and slip a latex sleeve over each

of them, as shown in Figure 13-8. Make sure the tubes are snug against the base of the hardboard and then secure them with zip ties. Snip off the dangling ends using scissors.

Figure 13-8 *Assembling the end effector*

6. Now get your jumbo-sized Popsicle sticks and drill a 5/16″ hole in one and a pair of 1/16″ holes in the other side using the PDF Popsicle template as a guide. These can be incredible fragile, so drill pilot holes first and then drill slowly on top of a piece of scrap wood or cardboard. Don't press too hard! See Figure 13-9.

Figure 13-9 *Drilling through Popsicle sticks*

7. Now slip a 1″ dowel through each 5/16″ hole and cap the ends with latex tubing, as shown in Figure 13-10. The fit should be snug on the sides, but the dowel should be able to rotate freely. If you have a hard time getting the latex over the dowel, use a little cornstarch in the end of the tube to reduce friction.

Figure 13-10 *Assembling upper arms*

8. Now slip the longer dowels into each side of the upper arm, making a T shape, as shown in Figure 13-11.

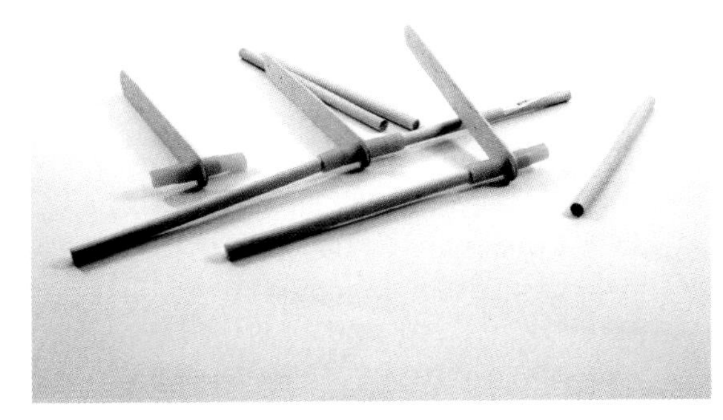

Figure 13-11 *Attaching lower arms*

9. Now bend the long arms 90° and slip them into the tubes of the end effector, as shown in Figure 13-12.

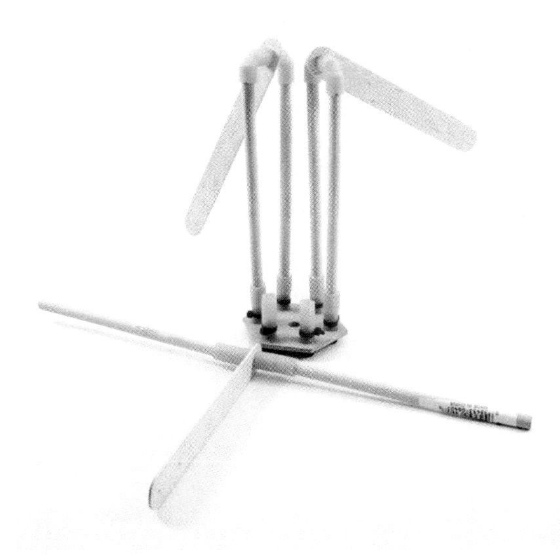

Figure 13-12 *Attaching the end effector*

10. As with the end effector, drill out the holes indicated on the top platform template (Figure 13-13).

Figure 13-13 *Drilling the top platform*

11. Now tape three pieces of cardboard together, and sandwich them between the servo and the top platform. Then zip-tie the whole thing to the top platform. This cardboard shim provides a little bit of clearance to make sure the top part of the servo arm can rotate freely. If you have something stronger than cardboard like a few extra bits of hardboard or other wood, that's even better! Just make sure you have about 1/2" of clearance. Pay attention to the orientation of the servo motors as indicated on the template. There was a little awkward space left over, so I've cut out a carrying handle. This is completely optional. Figure 13-14 shows the servos attached.

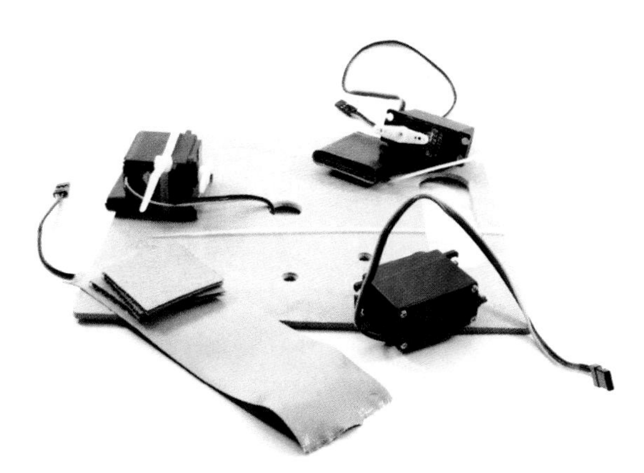

Figure 13-14 *Attaching servo motors*

12. Finally, using the small, double-sided servo horn on the servo, attach the Popsicle sticks to the servo horns with the #4 screws. To orient things correctly, make sure the long side of each servo is parallel to the Popsicle stick. You'll need to manually turn the servo all the way back until it stops with your fingers so that it can go from parallel to the base to fully vertical. If you can't extend the arm, pop the servo horn off, turn the servo back, and pop the horn back on until it has the correct range of motion. Now you're done assembling the mechanical parts. Your bot should look like Figure 13-15 at this point.

Figure 13-15 *Attaching the arms to the servo horns*

13. The wiring is relatively simple. In Figure 13-16, I use a breadboard, but it's almost overkill because you're really only using the power strip. You can connect the power and ground however you like, including using small wire nuts to hold everything together. Connect all of the red wires from the servo motors to the positive (+) terminal on your power source. Connect all of the black wires on the servo to the ground (-) terminal and then connect these to the GND pin on your Arduino to create a common ground. Finally, connect each signal wire (usually white or yellow) to a PWM pin on the Arduino. Here I used 9, 10, and 11. If you're not sure which pins are PWM capable on your Arduino, check the documentation for your specific model. That's all there is to it! Most of the complexity of a delta is in the software.

Making It Move

Connect your Arduino via USB, and try out the code from Example 13-1. Make sure to substitute your correct pin numbers for the servos if you've wired things up differently than in the wiring diagram. If you followed the wiring diagram precisely, you won't need to change the code.

All source code for the examples in this book can be found on GitHub (*https://github.com/rwaldron/javascript-robotics*).

Install Firmata first!

You'll want to make sure that your Arduino has the Firmata sketch installed before attempting to run any code. See "Arduino" for more details.

Example 13-1 *warmup.js (a simple code example to move the bot and make sure everything is working well)*

```
var five = require("johnny-five"),
  temporal = require("temporal"),
  board = new five.Board();

board.on("ready", function() {
  var servo1 = five.Servo({ pin:  9, range: [0,90] }),
      servo2 = five.Servo({ pin: 10, range: [0,90] }),
```

```
    servo3 = five.Servo({ pin: 11, range: [0,90] });

  var repeat = function() {
    temporal.queue([
      { delay: 250, task: function() { servo1.to(60); } },
      { delay: 250, task: function() { servo2.to(60); } },
      { delay: 250, task: function() { servo3.to(60); } },
      { delay: 250, task: function() { servo1.to(20); } },
      { delay: 250, task: function() { servo2.to(20); } },
      { delay: 250, task: function() { servo3.to(20); } },
      { delay: 250, task: repeat }
    ]);
  };
  repeat();
});
```

Here I am using the `temporal` npm library (see the Appendix for how to install npm modules) to sequentially move each arm. temporal's `queue` function takes a list of objects with a `delay` key and a `task` key. It will execute each of the tasks sequentially after the specified delay. By delaying longer than the amount of time it takes the servo to complete its movement, you can ensure that you're only moving one arm at a time. The last task in the list is a reference to the function itself so that you keep repeating the entire process forever and ever.

If everything worked as it should, when you run that code, you should see each arm of the delta move down 60° in sequence, and then move each arm back up to 20° before starting the sequence over again. If one or more of the arms doesn't work, check the wiring. Make sure each motor is connected properly to the batteries and the correct pins (9, 10, 11) on the Arduino.

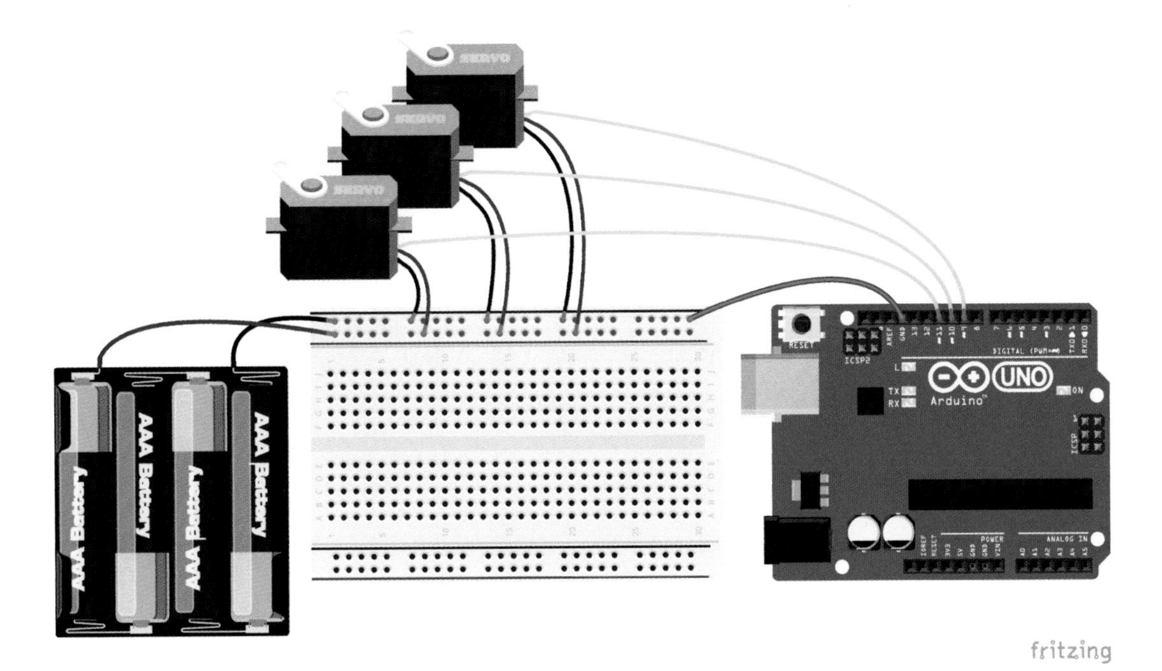

Figure 13-16 *Wiring diagram*

If a motor seems *stuck* or is clicking at you angrily, stop the program and check the range of motion on that arm by moving it back and forth manually. If it doesn't seem right, pop the servo horn off and twist the servo's hub into a better position, then try again until everything is moving smoothly.

If you watch the program run for a while, you'll get a sense for how the changing position of each arm affects the position of the end effector. Cool! It's not very useful though. In order to make your delta robot more useful, you'll need to find some way to determine the angles you need for each arm to put the end effector into the exact position you'd like it to be. This sounds like a job for mathematics.

Predictable Positioning Through Kinematics

Kinematics is a branch of mechanics dealing with the geometry of motion. In robotics, we are mostly concerned with mechanical systems made up of linked rigid bodies and joints, collectively called *kinematic chains*. For the purpose of this chapter, we don't have to get much deeper than that, other than to know that if you can describe the physical dimensions and constraints of a kinematic chain, you can predict the positions of the linked bodies mathematically as you manipulate the positions joints in a chain. Luckily, Mr. Clavel, the creator of the delta robot, has already done the hard work for you, and you need only provide some of physical constants of your robot to adapt the equations to your specific design. See Figure 13-17.

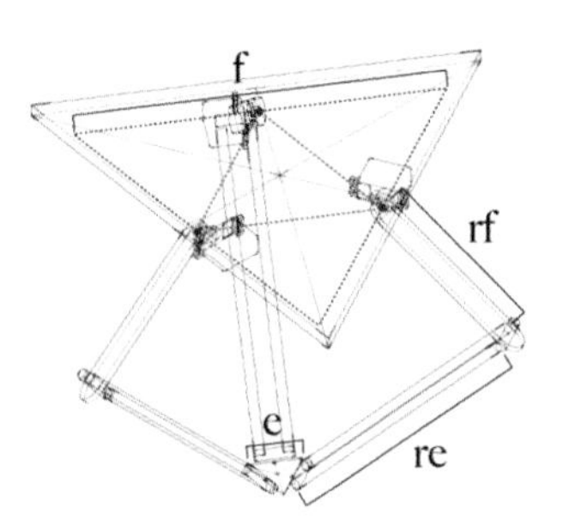

Figure 13-17 *Delta measurements required for the kinematic equation*

First, you need the measurement of one of the sides of the top triangle formed by the pivot points of the three motors. It's important that you measure from the pivot points and not the size of the platform they are attached to. You can also measure the diagonal distance between any two adjacent motors (designated by a dotted green line in the diagram) and double it. I'll call that number *f*. Next, you'll need the side length of end effector, measured the same way—the triangle formed by the pivot points. I'll call that number *e*. Finally, you need the length of the upper and lower arms. I'll use *rf* for the upper arm, and *re* for the lower arm.

```
var e = 80.25,
    f = 163,
    re = 155,
    rf = 128.75;
```

One of the beautiful things about the design of delta robots is that they are symmetrical in nature. You can use this symmetry to your advantage and simplify your problem so that you only have to look at one arm at a time. Let's take a look at a side view of your delta (Figure 13-18).

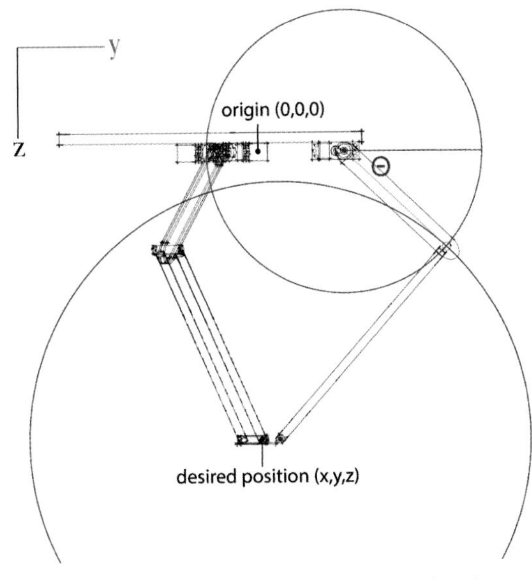

Figure 13-18 *Looking at the kinematics problem for a single arm*

The servo motor can only move in two directions on the *Y-Z* plane, so I'll just ignore the *X* dimension for the time being. The problem you are left with is the intersection of two circles: the first whose center is at the hub of your motor and has a radius of *f* (the length of your upper arm), and the second whose center is at the joint where the lower arm connects to the end effector and whose radius is *e* (the length of your lower arm). Where the circles intersect is where the upper/lower arm joint must be positioned for your end effector to be positioned where you want it. There are two intersection points, and you choose the outermost one so that your delta doesn't get too bent out of shape. The code for this is shown in Example 13-2.

Example 13-2 *Inverse kinematics: calculating angle theta for a single arm on the Y-Z plane*

```
// Calculates angle theta1 (for YZ-pane)
function delta_calcAngleYZ(x0, y0, z0) {
  var y1 = -0.5 * 0.57735 * f, // f/2 * tan(30 degrees)
      y0 -= 0.5 * 0.57735 * e; // Shift center to edge of effector

  // z = a + b*y
  var a = (x0 * x0 + y0 * y0 + z0 * z0
          + rf * rf - re * re - y1 * y1) / (2.0 * z0),
      b = (y1 - y0) / z0;

  // Discriminant
  var d = -(a + b * y1) * (a + b * y1)
           + rf * (b * b * rf + rf);
  if (d < 0) {
    // Non-existing position. return early with error.
    return [1, 0];
  }

  // Choose outer position of cicle
  var yj = (y1 - a * b - Math.sqrt(d)) / (b * b + 1);
  var zj = a + b * yj;
  var theta = Math.atan(-zj / (y1 - yj)) * 180.0
              / Math.PI + ((yj > y1) ? 180.0 : 0.0);

  return [0, theta]; // Return error, theta
};
```

That gets the position for one arm. Now to get to the next one: you just rotate your points 120° and run it again, and then rotate and run again for the third arm. None of the arms calculate the position of the *X* position directly, but as you rotate around and calculate *Y-Z* for each of the three angles, the end effector is pushed into the correct position and the universal joints between the end effector allow it to move smoothly and freely on those planes. Example 13-3 shows this code.

Example 13-3 *Calling delta_calcAngleYZ three times to position end effector*

```
// Calculate theta for each arm
function inverse(x0, y0, z0) {
  var theta1 = 0,
      theta2 = 0,
      theta3 = 0,
      cos120 = Math.cos(Math.PI * (120/180)),
      sin120 = Math.sin(Math.PI * (120/180)),
      status = delta_calcAngleYZ(x0, y0, z0);

  if (status[0] === 0) {
    theta1 = status[1];
    status = delta_calcAngleYZ(x0 * cos120 + y0 * sin120,
        y0 * cos120 - x0 * sin120, z0, theta2);
  }

  if (status[0] === 0) {
    theta2 = status[1];
    status = delta_calcAngleYZ(x0 * cos120 - y0 * sin120,
        y0 * cos120 + x0 * sin120, z0, theta3);
    theta3 = status[1];
  }

  return [status[0], theta1, theta2, theta3];
};
```

Once you can calculate angles for all three positions, you'll be able to position the end effector anywhere within your delta's physical range of motion. See Example 13-4.

Example 13-4 *Function to position the end effector at a given X, Y, and Z position*

```
var board = new five.Board();

board.on("ready", function() {

    // Setup
    var servo1 = five.Servo({
        pin: 9,
        range: [0,90]
    });
    var servo2 = five.Servo({
        pin: 10,
        range: [0,90]
    });
```

```
    var servo3 = five.Servo({
        pin: 11,
        range: [0, 90]
    });

    function go(x, y, z, ms) {
      var angles = inverse(x, y, z);
      servo1.to(angles[1], ms);
      servo2.to(angles[2], ms);
      servo3.to(angles[3], ms);
      console.log(angles);
    };

    board.repl.inject({
      go: go
    });

    // Initial position
    go(0,0,-150);

});
```

In the preceding code, I am defining a *go* function that will let you move the end effector wherever you like. The initial position is set to *X*=0, *Y*=0, *Z*=-150. The origin is set in the center of the top platform, but your end effector doesn't go that high because of the space that the motors and folded up arms take up! So *Z*=-150 is considered the highest *Z* position. Make the number lower to bring the arm down.

You can now control the delta robot from the REPL! Run the code from GitHub (*https://github.com/rwaldron/javascript-robotics/tree/master/delta*) and try typing the following:

```
go(0,0,-180); ❶

go(50,0,-160);  ❷
go(-50,0,-160); ❸

go(50,50,-160); ❹
```

❶ Move down

❷ 50 mm in the *X* axis

❸ Go to the other extreme of the *X* axis

❹ Move *X* and *Y*

Now you can bring temporal back into the mix and draw a simple box, shown in Example 13-5.

If you attach a pen to the end effector, you could even coax it into drawing that box on a sheet of paper—just set your *Z* position correctly so the pen touches the paper. Congratulations! You've made a delta robot that can do things! What will it do next? That part is up to you.

How Deep Does This Rabbit Hole Go?

If you're looking for an even more in-depth explanation of the math behind this, you might want to check out this great tutorial (*http://bit.ly/17Qtus9*) by Trossen Robotics user *mzavatsky*. Almost all of the other delta sources cite it, and this chapter is no exception. Further, if you *really* like math, see his source material, Prof. Paul Zsombor-Murray's "Descriptive Geometric Kinematic Analysis of Clavel's *Delta* Robot" (PDF) (*http://www.cim.mcgill.ca/~paul/clavdelt.pdf*). I'd also like to give special thanks to Jason Huggins, whose delta was the first I played with, and whose JavaScript source code was adapted for this chapter.

Example 13-5 *Drawing a box*

```
function box() {
  temporal.queue([
    { delay: 250, task: function() { go( 30,  30, -160, 250); } },
    { delay: 250, task: function() { go( 30, -30, -160, 250); } },
    { delay: 250, task: function() { go(-30, -30, -160, 250); } },
    { delay: 250, task: function() { go(-30,  30, -160, 250); } },
    { delay: 250, task: function() { go( 30,  30, -160, 250); } }
  ]);
}

board.repl.inject({
  box: box
});
```

More Sophisticated Delta Options

Junky Delta was designed to be simple and cheap to put together, but maybe you want a nicer option? The good news is that if you have access to a 3D printer, there are a couple of really nice kits you can print instead. All of the code and lessons we just reviewed still apply, but you'll get additional points for style and convenience!

TapsterBot

TapsterBot (*https://github.com/hugs/tapster bot*), shown in Figure 13-19, is a well-known, open source (BSD licensed) Delta Robot designed by Jason Huggins for mobile device testing. It's based on a previous design of his called BitBeam Bot, which was a traditional belt-driven, Cartesian device. Switching to a delta design allowed him to improve the speed, accuracy, and simplicity of the design.

Figure 13-19 *Jason Huggins' TapsterBot sporting an unapproved color scheme*

TapsterBot 1 is great design to start with if you own a 3D printer. Most of the parts are 3D printable, and all that remains to purchase is a slew of nuts and bolts and a few servo motors. TapsterBot 2 uses magnetic joints and has a few more exotic parts to track down.

Assembly is fairly straightforward, and Flickr user abbyraskin has a great step-by-step guide (*http://bit.ly/19M8GT6*).

Robot Army

The Robot Army delta (*http://robot-army.com*), shown in Figure 13-20, was successfully funded as a Kickstarter project to make a very inexpen-

Programming the Official Robot Army kit

If you have an official Robot Army kit, you'll need an FTDI USB cable to interface with the microcontroller, which is a custom variant of an Arduino Pro. You can get one at SparkFun, or just connect up a regular Arduino instead.

sive delta robot. During the kickstarter, they were selling for $100. Why are they so cheap? Because the team behind the project is making hundreds of them for an art installation. It's also interesting to note that this is an inverted delta —the base is on the bottom and the end effector extends upward. The tool on the end effector is an RGB LED that can be used for pretty, interactive light displays and artsy things like on the cover of this book.

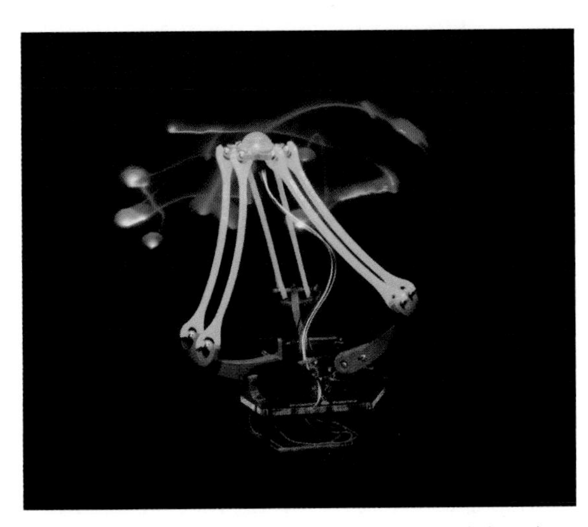

Figure 13-20 *Robot Army's Delta Robot is stylish and fun!*

You can buy a kit from them, or you can download and 3D print the parts yourself from their site. You won't have the custom electronics, but a standard Arduino will work just as well. They have a beautiful, Ikea-style assembly guide posted on their site (*http://robot-army.com/deltaInstructionsREV_C.pdf*).

What's Next?

Now you've got the basic knowledge to build and design a delta robot and make it move to a desired point on the *X*, *Y*, and *Z* axis. What will you do next? Here are some ideas:

- Slip a pen or brush into the delta's hand and teach it to draw!
- Combine with the OpenCV (*http://opencv.org*) and create a pick-and-place to separate M&M's, Skittles, and Reese's Pieces from each other by color.
- Build a larger version with a mechanical gripper arm and use it to pick and place kittens (for example)!
- Using inverse kinematics alone with servos isn't very accurate, especially because the servo motors can only move to certain angles. You can look into forward kinematics (where you input angles and get an *X*,*Y*,*Z* position) and compare to see how close you got and adjust for error correction.

The only limit is your imagination. Treat your delta robot well and you will have a loyal friend and lifelong companion. Enjoy!

14 Meow Shoes

By Suz Hinton

Have you ever wanted to turn your body into a device to communicate with computers, beyond just typing on a keyboard or clicking with a mouse? I bet you have. Expressing oneself through human movement is a deeply connective and fun experience for almost everyone. Dance and similar art forms can move beyond just the visual and auditory effects bound to the movements themselves.

What if you could augment this behavior into electronic communication? You can certainly do this with the help of JavaScript, and some rudimentary sensors placed in an otherwise unassuming pair of shoes! Are you ready to create magic with your feet?

Figure 14-1 *The finished Meow Shoes*

Bill of Materials

You will need the items listed in Table 14-1.

Table 14-1 *List of parts needed for Meow Shoes*

Part/item	Notes	Source	Estimated price
Arduino	For this project, the Arduino Micro (headerless version) is recommended for its tiny size, but use an Arduino Uno if you're new	AF 1315	$35 (approximately)

Part/item	Notes	Source	Estimated price
	to this stuff (and skip the soldering!)		
Four large FSR sensors	Square or round shapes are both good	AF 1075, SF SEN-09376	$7.95
Ribbon cable or hookup wire	Some soldering is required	AF 289, SF PRT-08024, MS MKEE3	$2.50
Four resistors	10k Ohm	SF COM-11508	$1
Soldering iron	Any old one will do	AF 180	$15–$25
Solder	The lead type is easier for beginners, but you may prefer lead-free	AF 145	$5.95
Wire strippers	These are so handy to have	AF 527	$11.95
JST connectors with wire	You'll need three male/female pairs	AF 578	$1.50 per pair
Stanley knife/sharp scissors	Don't run with them		
Electrical tape	For insulating	Hardware stores	$5
Glue gun/Superglue	For, erm, gluing	Hardware stores	$5–$10
Small piece of velcro	For sticking the Arduino to the shoes	Craft, hobby, or hardware stores	$5
Coiled microphone wire	Approx 0.5 m in length	eBay is a cheap place to find this	$12
A micro USB cable	The longer the better!	*http://bit.ly/19M8KSW*	A few dollars
A pair of shoes	To sacrifice		
A pair of shoe insoles	Either the fabric or the gel type	Drug stores/ pharmacy	$7
3D printer	Optional, to print helpful parts		

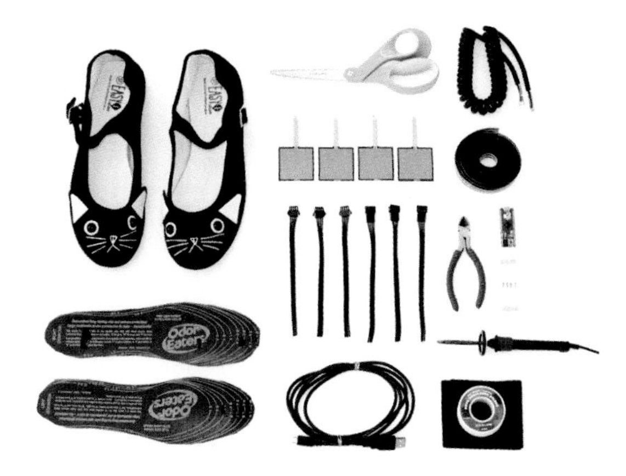

Figure 14-2 *Parts needed*

Parts Explained

Some of these required parts seem a little complicated, so what do they do?

Arduino

An Arduino is an electronic microcontroller board, with an easy to use development ecosystem. The included bootloader on the chip makes it very easy to upload programs, or sketches as they're called, from a computer to the board. The Arduino was designed to allow interaction with other hardware (e.g., sensors and mechanical parts). You can use it to read data from, talk to, and control hardware! It does this by featuring both digital and analog pins, which can read and write data to a large range of devices. We'll be using the analog pins in the project, to read data from some pressure sensors.

Force-sensitive resistor (FSR)

A force-sensitive resistor (or FSR for short) is a type of analog sensor that can measure the amount of pressure or force being applied to it. It outputs this resistance value in Ohms.

So how does it work? An FSR is a rather simple piece of technology when you take it apart. The sensor consists of three layers: the active area, the spacer, and the conductive layer. The active area layer has a conducive path that snakes its way back and forth along the surface. This creates some of the current resistance when force is applied. The conductive layer is named so for the large pad of conductive surface material present. Current will also pass through this area. The spacer layer simply keeps the two conductive layers separate from each other when force is not present. This essentially creates an open circuit when no force is being applied.

What happens to these layers when force is applied? The sensor condenses or squashes the layers together. When current is passed through the sensor as this is happening, the amount of resistance created by the top and bottom layers touching each other (remember the snake path on the active area?) changes. If you apply heavy and even force to the entire sensor area, you'll see a lower resistance happen. If you only press on a small area, or press lightly, you'll see a higher resistance. Pretty logical, right? Figure 14-3 shows the layers.

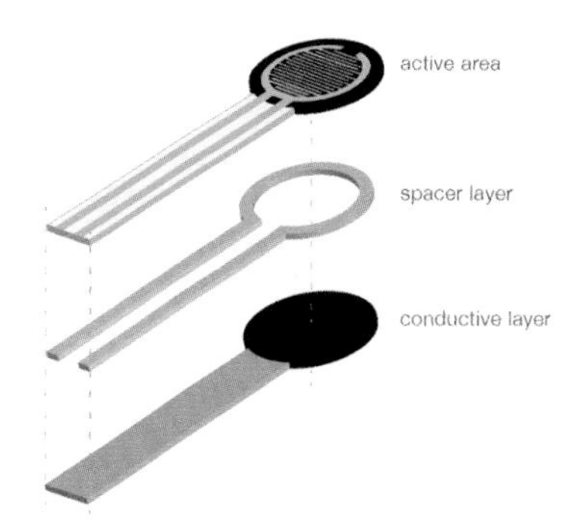

Figure 14-3 *Cross section of a force-sensitive resistor*

One thing to keep in mind with FSRs—the resistance change is not linear as pressure is applied. Resistance lowers in a much more dramatic fashion when applying light pressure, then steadies out a little more from there as the source of pressure increases. Figure 14-4 shows the graph of the resistive change.

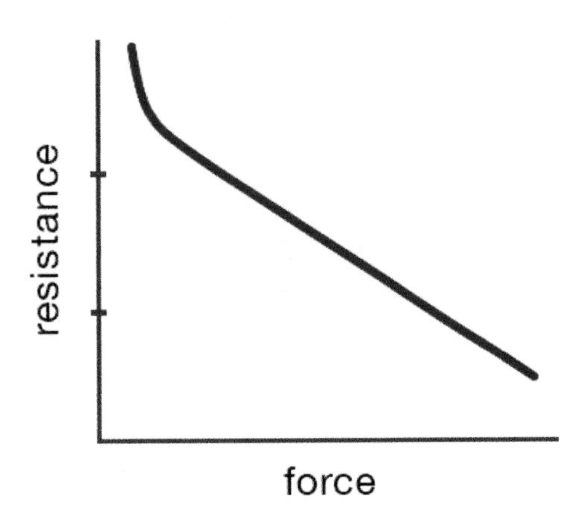

Figure 14-4 *Graph visualizing resistive change as nonlinear*

10K Ohm resistor
A resistor is an electronic component designed to reduce voltage levels in a circuit. It does this by burning off energy from the supplied current as either light or heat. An Ohm is the unit of measurement we use to measure resistance.

The topic of resistors is a very in-depth and mathematical one (we encourage you to read up—it's complex, but good to understand). To keep things simple in the context of this particular project, we are simply using resistors to pull down the logical low value reading for our sensors. If the circuit is open (no force being applied), the analog reading we will get from our Arduino analog pin will be a more constant 0, thanks to the pull down behavior of our resistor. No force present equalling 0 in our data readings is not only logical, but this value will be a little more reliable without the fluctuations in the circuit we'd normally see affecting the sensor data.

Micro USB cable
We all know what a USB cable is, but how are we using it in this project? You'd be right if you assumed it would connect the Arduino to your computer of choice. The sensor data will be sent from the Arduino over USB, eventually being read from serial into our JavaScript program via "WebSockets." Cool!

3D printer
3D printers are relatively easy-to-use machines. They are used to manufacture plastic, metal, or ceramic parts. In this project, having a run-of-the-mill plastic 3D printer is a bonus. We can print a case for our Arduino to make it look fancy and professional. This will also protect some of the fragile wiring and soldering you spent so much time on! Don't have a 3D printer? Hit up your local hackerspace if you have one in town. We even hear that UPS is starting to offer 3D printing as a service. Alternatively, the helpful folks at Shapeways (*http://shapeways.com*) or 3D Hubs (*http://3dhubs.com*) will take an uploaded model and send you the print in a week or so. Neat!

Making the Sensor Inserts

The first task to do is to prepare the force sensors for your shoes!

There are a few skills involved in this section:

- Soldering
- Laying the wiring as flat as possible
- Getting the sensor placement correct

Solder the Sensors to the Wiring

Before you install the sensors in each shoe, you'll need to do a little soldering first. Each sensor will need two wires connected to the terminals:

1. Take one of the female JST cables, and cut two of the wires down really short. Strip the ends of each wire, to prepare them for soldering, as shown in Figure 14-5.

Figure 14-5 *Wires cut short*

2. Solder the terminals of a sensor to one pair of wires (Figure 14-6). Then repeat for the other sensor (Figure 14-7).

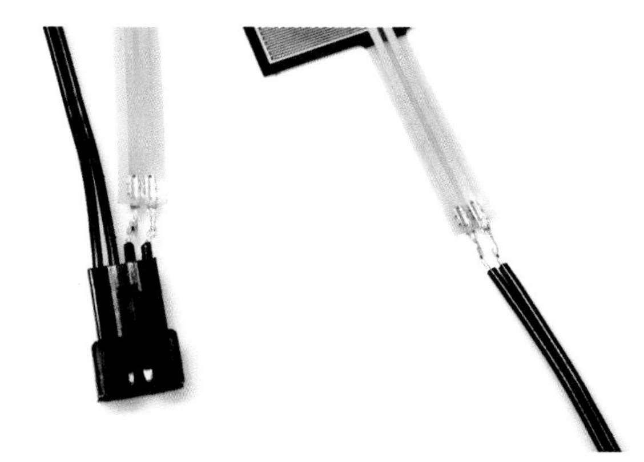

Figure 14-6 *How each pair of wires should look*

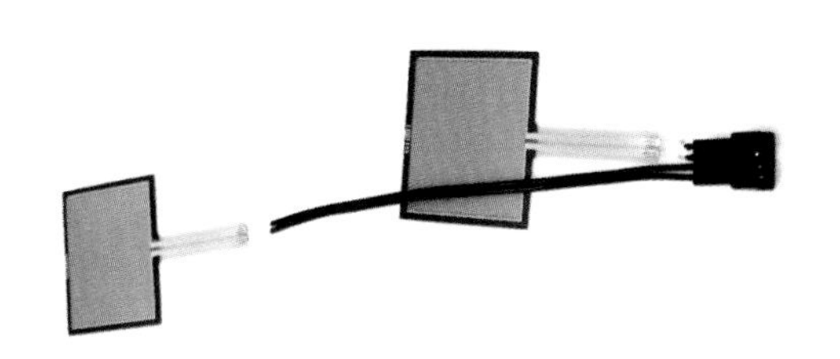

Figure 14-7 *The final outcome*

3. Wrap each individual connection with electrical tape to help protect the soldered connections, as shown in Figure 14-8.

Figure 14-8 *Electrical tape insulates and cushions force applied*

Install Sensors into the Shoes

OK, now you're ready to attach the sensors to the shoes. Each sensor has a sticky backing that can be peeled off. Use this to stick them to the footbed of the shoes:

1. Stick the sensor with the longer wire pair to the footbed, where the ball of your foot would rest. The wire should be facing toward the back of the shoe.
2. Stick the sensor with the shorter wire pair to the footbed, where your heel would rest.

Tunnel the Wiring out of the Shoe

Once you're happy with the placement, it's time to route the wires out of the shoe:

1. Cut a hole in the back of the shoe right down the bottom where the sole begins, as shown in Figure 14-9.

Figure 14-9 *The hole above is large enough to fit the female JST plug*

1. Thread the female connector out through the hole, with the sensors still inside the shoe.
2. Tuck the connector partly into the shoe hole and glue in place, as shown in Figure 14-10.

Figure 14-10 *Final assembly of female JST wire within shoe*

Repeat the preceding steps for the other shoe.

Connecting the Shoes

In order for both shoes to talk to the Arduino, the most practical method is to connect the two shoes in a way that won't hamper their use. You'll use the coiled, stretchy microphone cord to allow more natural movement. You're going to use male JST connectors this time, so that the wire just clicks in and out of the shoes to connect them neatly.

The following directions assume you have no experience in wire crimping, which is a better way to join wires to connectors.

Normally you would disassemble the JST connectors, remove the black wires, and crimp the microphone cord wires to the terminals.

If you're experienced in crimping, go for it! Otherwise, you'll be soldering wire-to-wire if you're less experienced with this stuff. Everything will still work the same.

Prepare the Coiled Connector Cord

To prepare the coiled connected cord, follow these steps:

1. Cut the wires short on two male connectors, and strip the ends (see Figure 14-11).

Figure 14-11 *All four wires are cut very short*

2. Choose four wire colors to use in the microphone cord. Cut the others away. Solder the four wires of each end of the cord to a male connector. The connec-

tor cords will need to be cut short and stripped first.

Make sure you solder with the exact same wiring order for each connector end. Figure 14-12 shows this.

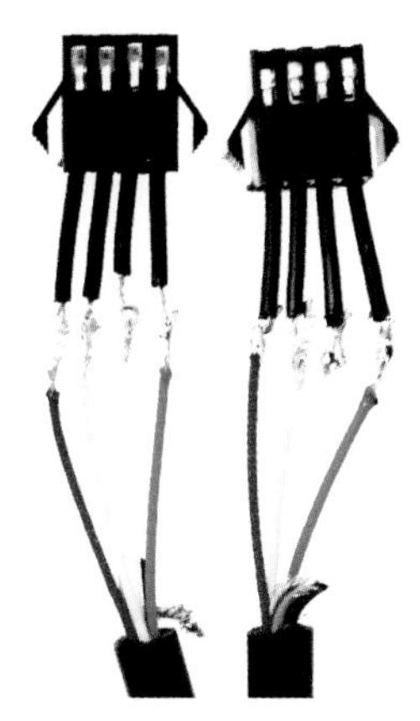

Figure 14-12 *All four chosen wires soldered, the others cut away completely*

3. Wrap each connection with tape to help protect the soldered connections. Then wrap electrical tape around the whole assembly, sealing it all off neatly (Figure 14-13).

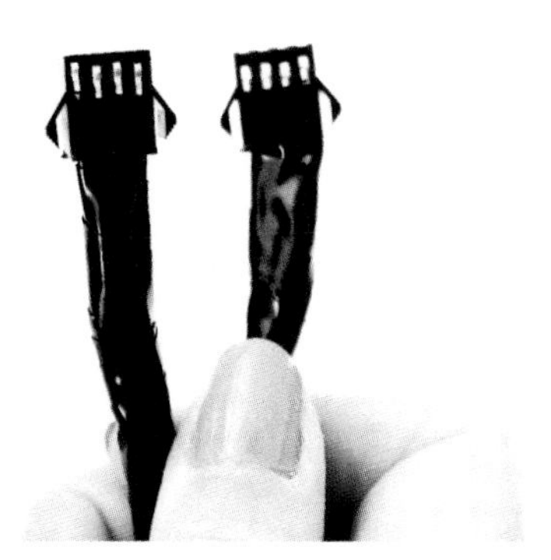

Figure 14-13 *Make it nice and neat, with no exposed wiring*

4. Cut two slits in the right shoe, above the existing connector hole.

5. Thread a female JST connector cable in and out, then secure the connector to the shoe with glue, as shown in Figure 14-14.

Figure 14-14 *Final right shoe assembly*

6. Click in each male connector to its female pairing to join the shoes together (Figure 14-15).

Figure 14-15 *Your shoes should now look something like this picture*

Nice! Insert the insoles so that they're laying over the sensors as a top layer. This will help protect them from moisture and stampy feet!

Connect the Shoes to an Arduino

You shoes need something to talk to, right? That's where the Arduino comes in. How are FSRs connected to an Arduino?

First, let's look at an example of just one FSR, shown in the Fritzing diagram (*http://fritzing.org*) (Figure 14-16).

One sensor terminal goes to 5V. The other, to both ground and an analog pin. In this example, A0. The pull-down resister mentioned earlier in this project exists between the ground and the analog pin. What does it look like when all four sensors are connected? Like a mess! So let's look at it when using a breadboard (Figure 14-17).

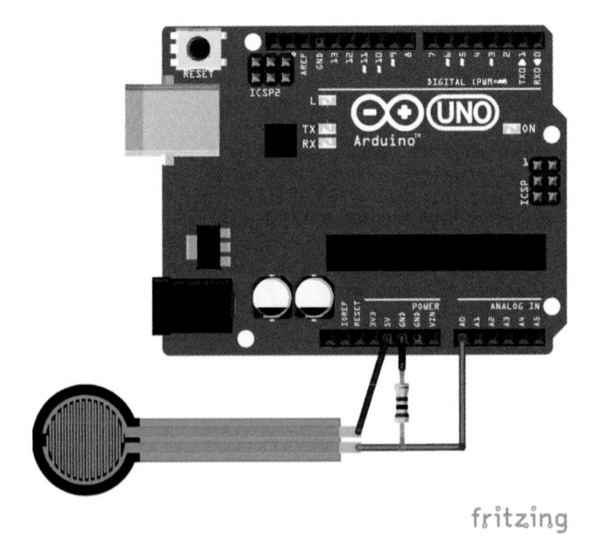

Figure 14-16 *How to connect one FSR to an Arduino*

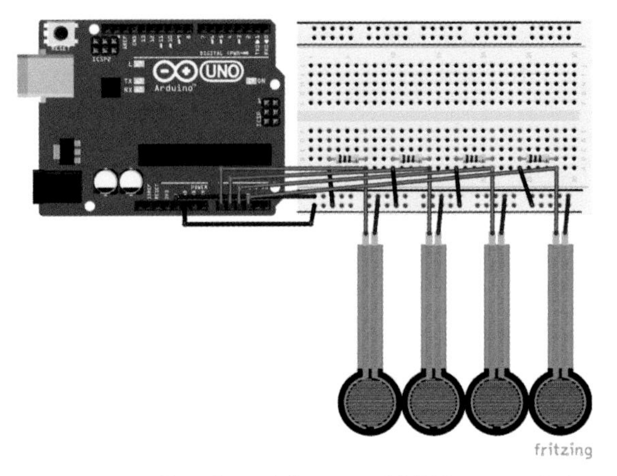

Figure 14-17 *Breadboard diagram of 4 FSRs connected to an Arduino*

Each shoe has a JST connector, ready for the Arduino. Both are on the right shoe, which means you'll be mounting the Arduino to it.

Let's do one shoe at a time, and take it slowly.

Using an Arduino Uno? Just use the headers and optionally a mini breadboard instead of soldering in the following steps. Using an Arduino Micro? You'll be soldering. Yay!

Prepare Wiring

Follow these steps to prepare the wiring:

1. Take a male connector cable, and plug it into one of the right shoe female ports.
2. Split off the wires slightly, and strip them.
3. Solder a resistor to the second wire from the left.
4. Solder a short wire to the other end of that resistor.
5. Solder another resistor to the fourth wire from the left.

6. Solder a short wire to the other end of that resistor.

Arduino Soldering

Next, follow these steps:

1. Solder the resistor arms to the GND pin on the Arduino.
2. Solder the short wire coming from the resistor to its own Analog pin, starting with A0.
3. Solder the remaining wires to the 5V pin.

One shoe done! Now do the other one in the same way.

Attach the Arduino to the Right Shoe

While you're wearing your (almost finished!) sensor shoes, you'll need a safe and secure spot for your Arduino to hang out. This is where your Velcro comes in handy. We're going to attach the Arduino to the outer side of the right shoe.

If you 3D printed the Arduino Micro case, or found alternative housing for it (try using the cardboard box it came in!), place the Arduino in there first, and close the lid. Otherwise, stick the Velcro right onto the back of the Arduino Micro/Uno itself!

1. Peel the backing from one side of the Velcro (loop and sides both together).
2. Stick the Velcro to the bottom of the Arduino case/Arduino itself, as shown in Figure 14-18.

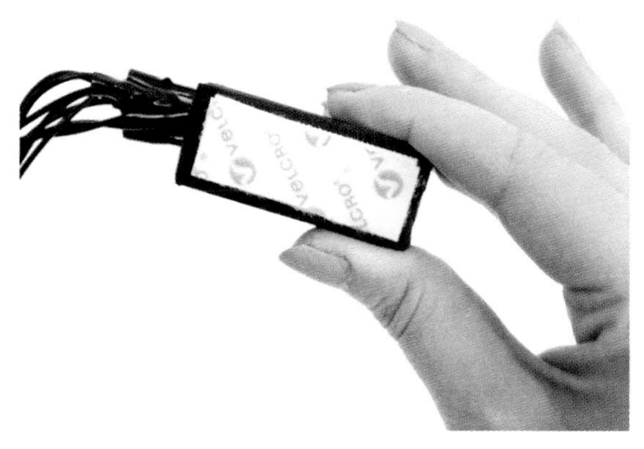

Figure 14-18 *Velcro attached to bottom of the case*

3. Peel the backing from the other side of the Velcro, and stick to the outside of the right shoe, as shown in Figure 14-19.

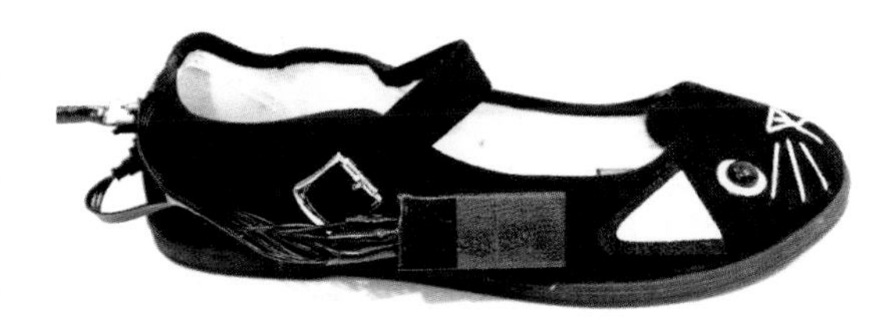

Figure 14-19 *The Arduino now attached to the side of the right shoe*

Your Meow Shoes are assembled! Let's get coding!

Running the Code with Johnny-Five

The Johnny-Five robotic library has out-of-the-box support for force-sensitive resistors, so you're in luck! Let's review how to get a simple test up and running.

Connecting to Johnny-Five

Before you type anything, plug the micro USB cable into the Arduino at one end, and the computer you're running the code on at the other end. This step is pretty obvious, but I've been burned in the past because the wrong identical looking USB cable was plugged in instead, doh!

If you haven't already done so, install Node.js, followed by the latest version of Johnny-Five from npm. Be sure you have the latest version of StandardFirmata running on your Arduino (see "Arduino"). All source code for the examples in this book can be found on GitHub (*https://github.com/rwaldron/javascript-robotics*). To install Node.js and Johnny-Five, run the following:

```
npm install johnny-five;
```

Once everything's installed, you should first do a quick sanity check that the computer and Arduino are getting along fine, and communicating in the same language so to speak.

The following code simply requires Johnny-Five, and instantiates your Arduino as a *board*. Once the board is all ready to go, you'll just log a message to the console/terminal.

For each short example shown here and in upcoming sections, you can save the code to a file, then type `node filename.js` *into your terminal to run it (be sure to replace "filename" with the name of the file you saved the code into).*

```
var five = require("johnny-five");
var board = new five.Board();

board.on("ready", function() {
  console.log("Meow Shoes say hello!");
});
```

The preceding code example creates a reference to the Johnny-Five library as *five*. It then creates a new variable *board*, and instantiates a new Arduino board with it. That board object will emit a *ready* event when Johnny-Five has successfully connected to the Arduino.

Did that work? Excellent! Onto the next task.

Setting up Sensors

Within your Johnny-Five code, you need to specify how many sensors you have (four), and what analog pins they are on. If you can think back to the assembly of the wiring, you'll remember that you connected your data wires to A0, A1, A2, and A3.

Only you know which pin you picked for which sensor, so either inspect the wiring and trace it back to each sensor to find out, or just guess and check until you get the following code referencing each correctly.

The following code will set up and name each sensor to a descriptive variable for easy referencing later. You'll need to specify the correct analog pin for each, and the frequency at which the data will emit:

```
var five = require("johnny-five");
var board = new five.Board();

board.on("ready", function() {
  var leftToe = new five.Sensor("A0");
  var leftHeel = new five.Sensor("A1");
  var rightToe = new five.Sensor("A2");
  var rightHeel = new five.Sensor("A3");
});
```

So what's happening here? Well, you need a new variable for each sensor, in order to track their data output separately. Johnny-Five has a class called `Sensor`, which we use when we instantiate a new copy of each FSR (two for each shoe). We're naming each variable descriptive names so things don't get too confusing!

Logging Output of Sensors

Now that your sensors have been set up, we gotta get some of that data out! This is pretty simple. You'll use the `data` event that is emitted every time the computer receives data from the Arduino via the USB serial connection.

In the following example, we're setting up a callback to console log the data value from each sensor (this allows you to make sure your wire connections are correctly set up):

```
leftToe.on("data", function() {
  console.log("left toe: ", this.value);
});

leftHeel.on("data", function() {
  console.log("left heel: ", this.value);
});

rightToe.on("data", function() {
  console.log("right toe: ", this.value);
});

rightHeel.on("data", function() {
  console.log("right heel: ", this.value);
});
```

What's this `data` event all about? Well, each time Johnny-Five receives some data over serial from the analog pins you connected your sensors to, the sensor values are sent in the data event emission. This event is simply catching this, and console-logging it out to the terminal.

When you first run this file via Node.js, you should see lots of logs spilling into your terminal! This is good, as it means that things are starting to work.

However, be sure to check for the following:

- All four sensor labels are showing up in your console log.
- All values should read 0 if no pressure is being applied to the sensors.

Double check your wiring if these sanity checks are not as expected.

If it's good to go, put the shoes on and start pressing each sensor one by one, jumping in the air, and dancing like you just made some amazing magic shoes. Look at those values change!

These values should fall between 0 and 1023. This will be the range you can play with when deciding on behaviors from your shoes to code for. Take a note of what the average value seems to be when your own body weight is applied to the shoes. A good ballpark to sanity check with is that medium pressure normally returns a value above 650, and heavy pressure is in the 900 figures, close to the 1023 limit.

Example Behavior

Let's look at setting up simple behaviors to perform actions in your code with.

For example, how would you check to see if someone wearing these shoes is standing still? Let's break this down first by imagining someone standing still, with both feet flat to the ground. Logically (and pretty obviously), both the toe and heel sensors belonging to a shoe would be having a consistent pressure applied to them, right?

Knowing this, you can apply simple conditionals to test for standing behavior. See the code loop in Example 14-1 (place it within your board's *ready* callback) which tests for a left shoe stand.

First, you're setting up two handy functions in the preceding example. The `isPressed()` function allows you to pass in a sensor data value, and it will then run a check to see if the value is high enough to validate a solid press of the sensor.

The second function, `isStanding()`, uses the `isPressed()` function declared earlier to test if all sensors are being pressed simultaneously. If so, it simply returns true. If not, it will return false. You can use this function as a boolean value in your code.

Pretty cool, huh? Now change this code to also report if you're standing on your right foot, and then test for standing on both feet!

You can use similar conditionals to test for things like heel and toe tapping. This is your next challenge, once you've mastered standing!

Example 14-1 *Testing for a left shoe stand*

```
// change this value to suit your weight/pressure needs, as explained above
var pressureThresh = 800;

// this tests if a sensor is currently being pressed hard enough
function isPressed(val) {
  if (val > pressureThresh) {
    return true;
  } else {
    return false;
  }
}

// this tests if both sensors in the left foot are being pressed simultaneously
function isStanding() {
  if (isPressed(leftToe.value) && isPressed(leftHeel.value)) {
    return true;
  } else {
    return false;
  }
}
// main loop, every 25 ms
this.loop(25, function() {
  if (isStanding()) {
    console.log("Standing on your left foot!");
  } else {
    console.log("Not standing");
  }
}); // end loop
```

What's Next?

You can do so many things with four simple pressure sensors. Here are a few things to get you thinking, I'm sure you will think of even cooler uses that will truly delight!

- Make each tap emit a different meow sound effect. Your pets and family might not be all that impressed with this as the novelty wears off. Take it from someone who knows.
- Create a music sequencer, with a different note coupled to each sensor. I have included a simple example of this to get you up and running in the code repository for this book. Yay!
- Navigate through an RPG game with your feet.
- Send secret Morse code messages to your loved ones via the art of interpretive dance.
- Create a painting application that lets you make art with simple choreography.
- Make another pair for a friend and compete in interactive balancing, running, and sports matches.

But most of all, enjoy your new Meow Shoes, and have fun coding!

Appendix

All of the projects in this book have prerequisites for your development environment and hardware configuration. For most projects, at a minimum, you'll need to install the latest stable version of Node.js, the Johnny-Five library, and StandardFirmata firmware on your Arduino. Projects that require a different setup will make those requirements clear and provide steps to get you up and running. If you need help installing and configuring your software or hardware, we've provided some basic instructions here to get you started.

Installing Node.js

To use Johnny-Five, you'll need Node.js v0.10.x or later. You can find prebuilt installers for Mac and Windows at *http://nodejs.org/download/*. On Linux, you should be able to install it from your Linux distribution's package manager (e.g., `apt-get install node` on Debian or Ubuntu). On Raspberry Pi, we suggest you use the packages from node-arm (*http://bit.ly/19M8Ny4*). If you need to update npm, you may do so using `npm install -g npm`.

Installing Johnny-Five

You can install Johnny-Five via *npm* (which comes with Node.js). In most cases, getting started is as simple as the following (on Mac or Linux):

```
mkdir nodebot && cd nodebot;
npm install johnny-five;
```

Now open your text editor and create a new file called *blink.js*. In that file, type or paste the following:

```
var five = require("johnny-five");
var board = new five.Board();

board.on("ready", function() {
  var led = new five.Led(13);

  led.blink();
});
```

Make sure the board you're using (usually an Arduino, which you must flash as described in "Arduino") is plugged into your host machine (desktop, laptop, or Raspberry Pi). In your terminal, type or paste the following:

```
node blink.js
```

The built-in LED on pin 13 should start blinking!

Troubleshooting

If the preceding instructions didn't work as expected, make sure that StandardFirmata is installed on the board (see "Arduino").

Sometimes Windows systems will fail to compile native dependencies, so if you run across this case, try the following:

```
npm install johnny-five --msvs_version=2012
```

More Information

The Johnny-Five Wiki has lots more information to get you started using the library, including a complete Getting Started Guide (*http://bit.ly/1BuTQH7*).

Configuring Your Hardware

Depending on which hardware platform you're using, you may need to do a little setup. Here are instructions to get you started on the platforms used by projects in this book.

Arduino

When using Johnny-Five and an Arduino, you'll need to be sure that the Firmata firmware is installed. Though some Arduino boards may come pre-flashed with a version of the Firmata firmware, you should make sure you're running the latest version. Follow these steps to install StandardFirmata on your board:

1. Download (*http://bit.ly/1lKHbJ4*) and install (*http://bit.ly/lGAqel*) the Arduino IDE.
2. Plug in your Arduino or Arduino compatible microcontroller via USB.
3. Open the Arduino IDE, select: File→Examples→Firmata→StandardFirmata.
4. Under the Tools menu, ensure the correct board type and serial port are selected.
5. Click the Upload button.

If the upload was successful, the board is now prepared and you can close the Arduino IDE. If you get an error, be sure the correct Arduino board type is selected (Tools→Board).

BeagleBone Black

Node should be preinstalled on your BeagleBone Black—however, it may be out of date. While connected to the Internet, run one of the following update commands:

Rev C/Debian

```
sudo apt-get update
sudo apt-get upgrade
```

Older models/Angstrom

```
sudo opkg update
sudo opkg upgrade
```

Raspberry Pi

It is best to start with a clean build of the Raspberry Pi. You can use NOOBS (*http://bit.ly/rasp-dl*); for help with installation, try Raspberry Pi's help section (*http://bit.ly/19M8YJX*). Be sure you have backed up any files on your SD card that you want to keep prior to formatting the card and installing NOOBS.

1. Download NOOBS from *http://www.rapberrypi.org/downloads/* and select the NOOBS ZIP download or torrent if you have a torrent client. This may take a long time to download.
2. Format your SD card before copying the NOOBS files onto it:
   - Visit *https://www.sdcard.org/downloads/formatter_4/* and download SD Formatter 4.0 for either Windows or Mac.
   - Install the software.
   - Insert the SD card into your computer's card reader and note the drive letter assigned to it (e.g. E:/).
   - In SD Formatter, select the drive letter for your SD card and format it.
3. When the download has finished, extract the files from the ZIP.
4. Copy the contents of the folder to your SD card, and then remove the card from your card reader and insert it into the Raspberry SD holder.
5. Connect your Raspberry Pi to your monitor or TV, starting with the hub and then connecting the mouse, the keyboard, and the WiFi adapter.
6. Switch the monitor/TV to the input that the Raspberry Pi is connected to.
7. Plug in your USB power lead, and you should see the NOOBS selection screen.
8. Select Raspbian—this part can take about 20 minutes.

When the install is finished, press the Enter key. Your Raspberry Pi will reboot and the Raspberry Pi Software Configuration tool will be displayed.

Configuring a WiFi adapter

Here is how to configure your WiFi adapter to connect to a network:

1. Enter the following on the command line to start the GUI desktop if it doesn't start automatically (Raspberry Pi usually starts the GUI by default):

   ```
   startx
   ```

2. Double-click the WiFi config which brings up the wpa_gui window.
3. Select the "Manage Networks" tab, click the scan button, and the Scan results window will open.
4. Click the scan button and double-click your network from the list. A new window will open. In the field labeled PSK, enter your network password if required and click the Add button to save your network. The window will close.
5. Close the window and the wpa_gui window as well.
6. Double-click LXTermail icon, when the window loads type:

   ```
   sudo shutdown -r now
   ```

7. The Raspberry Pi will now reboot.
8. Log on using the default username *pi* and the password *raspberry*.
9. Attempt an outside connection:

   ```
   ping www.cnn.com
   ```

10. If the following output is displayed, you have no network connection, and will need to start back at step 1:

    ```
    ping: unknown host www.cnn.com
    ```

You have now configured your WiFi adapter to connect your home network to the Internet, and you have tested it. At this point, you can work with your Raspberry Pi headless, meaning with no keyboard, mouse, or screen connected so you access the Raspberry Pi via its network port, WiFi adapter, or the console. Your PC and Raspberry Pi must be on the same network and have the same subnet.

Spark WiFi Development Kit

The Spark Core and Spark Photon (*http://www.spark.io*) are open source, Arduino-compatible, and WiFi-enabled microcontrollers. If you run into any problems with installation and configuration, you can check out Spark's troubleshooting guide online (*http://docs.spark.io/troubleshooting*).

If you haven't already done so, you'll need to claim your Spark device.

Spark provides a mobile application for both Android and iOS that automates this process, but you can also do it from the command line using their Spark CLI tool (*http://docs.spark.io/cli/*) (see the documentation (*http://docs.spark.io/cli/*) for detailed instructions for using the command-line tool):

1. Plug your Spark into your computer via USB. Once it is connected, it should start flashing blue—this means it's in "listening mode" and is ready to be set up. This should always be the case for a new Spark, but if your Spark has been used before, you can hold the MODE button for 3 seconds (or 10 if you want to erase all previous WiFi connections).
2. Install and run Spark CLI by typing the following in your terminal:

   ```
   npm install -g spark-cli
   spark setup
   ```

3. Follow the prompts and create or log in to a Spark account.

4. Take note of the access token that you're given during this setup. You'll need this later!
5. Spark CLI will take you through the process of setting up your Spark for the local WiFi network. Enter the security information when asked. Once the WiFi credentials are accepted and your Spark connects, your Spark will begin "breathing cyan" to indicate this. This is a slow fading in and out of the light as opposed to a steady blinking.
6. Once it's in this stage, press Enter and your Spark should be successfully claimed!

The Spark Cloud provides you with two keys that you'll need to access your Spark device.

1. You will need your general account Access Token as well as the Device ID for the specific Spark device you are working with. If you didn't save it as part of the setup process earlier, the Access Token is available under settings of the online Spark.io Editor (*http://Spark.io*). The Device ID for each of your claimed devices is available in the device list panel.
2. You have the option to give your Spark a nickname for easier reference. When authenticating the Spark in your code, you can specify either the nickname or the full Device ID.
3. It's a good idea to access your Spark Device credentials as properties of `process.env` and keep them out of our source code. Access your credentials as environment variables by creating a file in your home directory called *.sparkrc* that contains the following:

   ```
   export SPARK_TOKEN="your spark token"
   export SPARK_DEVICE_ID="your device
   id"
   ```

4. Add the following to your *.rc* file of choice:

   ```
   source ~/.sparkrc
   ```

To set up the environment variables on Windows, see Spark's documentation on environment variables (http://bit.ly/1BuWqwK).

When working with Johnny-Five, instead of using the Spark Cloud, you'll be communicating with the Spark device locally using a custom firmware called VoodooSpark. VoodooSpark mimics the standard Spark firmware API and provides access to the Spark functionality via a local TCP connection:

1. With the Spark connected to a WiFi network, open the Spark.io Editor (*http://Spark.io*) and log in to the Spark Cloud. Download VoodooSpark (*http://voodoospark.me*), then copy and paste the entire contents of *voodoospark.cpp* into the Spark.io Editor window. Click "Verify" and then "Flash" to load VoodooSpark onto the Spark in exactly the same way you'd load any other Spark Application.
2. Alternatively, you can use the Spark CLI tool, which already has a precompiled build of VoodooSpark. With the Spark powered on and connected via USB, you can run the following command:

   ```
   spark flash $SPARK_DEVICE_ID voodoo
   ```

Android Development

The following instructions are required for Chapter 6.

Installing Android Studio

You can get the Android Studio from the Android Developer site (*http://bit.ly/19M90Bm*).

After installing Android Studio, make sure you have the following SDKs installed:

- Google Play Services
- Google Repository
- Android Wear

Creating an Android Project

First, create a project called `VoiceController`, as shown in Figure A-1.

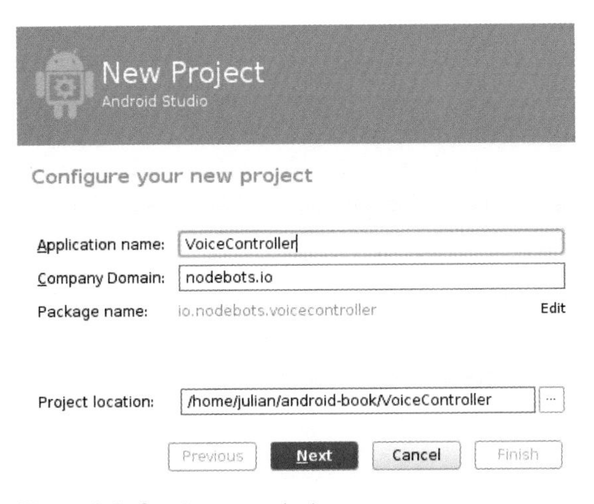

Figure A-1 *Create new project*

Then, select the SDKs used in this project. Because you are creating both mobile (phone and tablet) and wear applications, you will need to select both SDKs, as shown in Figure A-2.

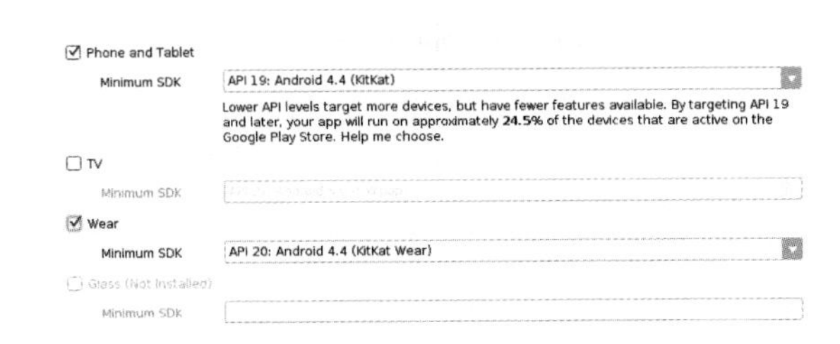

Figure A-2 *Select SDks*

Select a Blank Activity for your mobile application (Figure A-3).

Figure A-3 *Add Blank Activity to Mobile*

And create a MainActivity, as shown in Figure A-4.

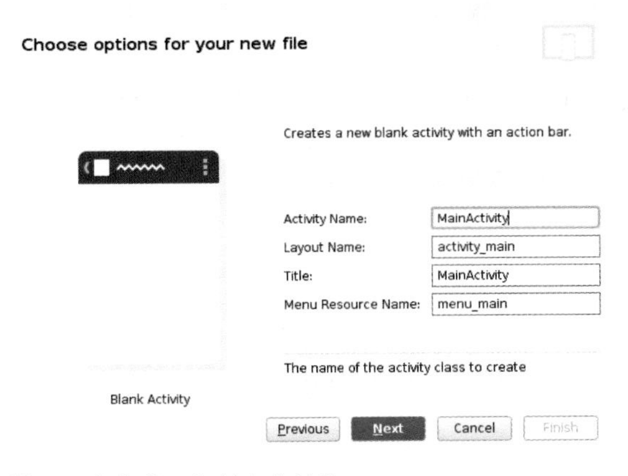

Figure A-4 *Create MainActivity*

Then add a Blank Wear Activity to your wear application, as shown in Figure A-5.

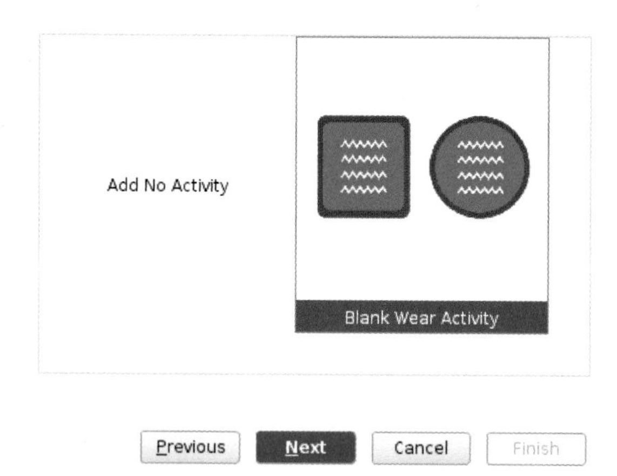

Figure A-5 *Add Blank Wear Activity*

And finally create a WearMainActivity (Figure A-6).

Figure A-6 *Create WearMainActivity*

Installing Volley

Download Volley (*http://bit.ly/1BTWibw*), then place the JAR in your *VoiceController/app/libs/* folder. In Android Studio, right-click it and select Add As Library.

Index

Symbols

A

B

C

D

E

M

N

O

P

R

S

T

U

V

W

X

Z

About the Authors

Jonathan Beri is a Maker of all sorts, with an affinity for APIs, robots, and obscure JavaScript frameworks. He currently works at Google as a Developer Advocate, making the lives of developers easier, one day at a time. As a Developer Advocate, he's helped launch the Google+ Hangout Apps Platform, Google+ Sign-In, and the Google Apps Marketplace. He can be found on Google+ as JonathanBeri.

Donovan Buck is an aspiring roboticist and a contributor to Johnny-Five—the open source, JavaScript, and Arduino programming framework. He is a strong believer in the value of continuing education in computer science. His own CS education began at age 5, ferrying punch cards around the data center for his dad and swapping out magnetic tapes while perched atop boxes of printer paper. He can be found on Twitter and GitHub as @dtex.

Kassandra Perch is an Open Web Engineer and Educator at Bocoup. She's been a JavaScript addict for her entire career as a programmer, and enjoys JavaScript in browsers, on the server, and of course on robots! When she's not trying new robotics platforms, she's working on crafting, or enjoying her evenings in Austin, Texas. She can be found on Twitter, GitHub, and most places as @nodebotanist.

Bryan Hughes is a frontend developer at Rdio and the lead organizer for the NodeBots SF meetup group. Bryan also created the Raspi-IO library which provides Raspberry Pi support for the Johnny-Five robotics framework. Bryan received his Ph.D. in Electrical Engineering and Computer Science from Texas Tech University in 2010. When not coding, he can be found spending time with his amazing partner and going wine tasting, attempting to become a photographer, or hiking. He can be found on Twitter at @nebrius and on GitHub as @bryan-m-hughes.

Raquel Vélez has been a core member of the NodeBots movement since 2012. Prior to becoming a web developer, she was a full-time roboticist, having studied Mechanical Engineering at the California Institute of Technology and completed some masters work in robotics engineering at the University of Genoa, Italy. She has worked at a variety of universities and laboratories around the globe, including the NASA Jet Propulsion Laboratory, the MIT Lincoln Laboratory, and Applied Minds, Inc. She can be found on Twitter and GitHub as @rockbot.

Lyza Danger Gardner is a dev. Since cofounding Portland, Oregon–based mobile web startup Cloud Four (*http://www.cloudfour.com*) in 2007, Lyza has tortured and thrilled herself with the intricate ins and outs of the bazillion devices and browsers now accessing the web globally. Lyza is also coauthor of *Head First Mobile Web* (O'Reilly). She can be found on Twitter and GitHub as @lyzadanger.

Susan Hinton is a JavaScript developer who likes to tinker with hardware. A Maker at heart, she's no stranger to minor burns from soldering irons and 3D printers. She's a regular contributor to the open source Node.js electronics scene, and enjoys teaching others how to immerse themselves in the nerdiverse. Suz can be found on Twitter and GitHub as @noopkat.

Rick Waldron is an Open Web Engineer at Bocoup and the creator of Johnny-Five, a JavaScript framework for hardware programming on the Node.js platform, and is working towards establishing standards for general hardware APIs. Currently supporting Arduino, BeagleBone, Raspberry Pi, Linino One, Pinoccio, Spark-Core, Light Blue Bean, pcDuino, Intel Galileo and Intel Edison, his work was recently highlighted at Intel's IDF2014. As a jQuery Core committer and former board member of the jQuery Foundation, Rick serves on Ecma TC39 as a representative of jQuery, channeling the project's vast real-world experience into contributions to the design of the next version of JavaScript. He can be found on both Twitter and GitHub as @rwaldron.

Sara Gorecki first discovered her love of code through hardware hacking and experimenting with Johnny-Five. An alum of both Cornell University and the Flatiron school, she's currently working as a Node Engineer at Penton Media. In her quest to strengthen her local JavaScript community she cofounded NYC's Queens JS meetup. She can be found on Twitter and GitHub as @opheliasdaisies.

Julián Duque is a developer and educator by passion currently working as software engineer at NodeSource. He organises multiple community events in Colombia like MedellinJS, NodeBots Day and JSConf Colombia. He loves sharing knowledge and is currently collaborating as evangelist in the io.js project and teaching programming fundamentals and JavaScript through NodeSchool and NodeBots events in Colombia and Latin America. He can be found on Twitter at @julian_duque and on GitHub as @julianduque.

David Resseguie began his programming journey at the age of 5 when his dad bought an Apple IIe computer. Instead of just playing Stickybear Numbers, he wanted to know how it worked. He's enjoyed making and developing things with technology ever since. David is an avid collector of toy robots and loves the opportunity to combine that interest with software development through hardware hacking, Internet of Things, and NodeBots. He also has a passion for STEAM education and enjoys speaking at schools, using robotics to get kids excited about careers in science and technology. He can be found on Twitter and GitHub as @Resseguie.

Andrew Fisher is a creator of things that combine web tech, physical computing, and lots of data. He is an interaction researcher and developer, exploring how behaviour influences and is influenced by technology and machines. Andrew is the lead organizer of NodeBots Melbourne and is a strong evangelist on the use of web technologies with hardware. He can be found on Twitter and GitHub @ajfisher.

Pawel Szymczykowski is a software engineer at Wedgies and an enthusiastic maker of various things both physical and code-based. He discovered his passion for both hardware and JavaScript as a result of the NodeBots movement at JSConf 2013. He came up with a simple open source laser cut sumo bot kit for the NodeBots Day event in Las Vegas which ended up making appearances at similar events around the world. You can find it online at *sumobotkit.com*. He can be found on Twitter @makenai.

Anna Gerber is a core member of NodeBotsAU. She is also a technical project manager with the ITEE eResearch Group at the University of Queensland, Australia. She can be found on Twitter @annagerber.

Emily Rose is a transhumanist with a passion for queer cyborg artistry. They are currently experimenting with ambient intelligence, adaptive automation, and evolving interfaces. Emily is a world-class speaker who has brought humor and enlightenment to audiences across the globe. Original NodeBots curator, founder of DanceJS, and one of the most interesting individuals in the known universe; Emily is an unnatural force of pure unbridled creativity. They can be reached via Twitter (*https://twitter.com/nexxylove*) or via GitHub (*https://github.com/emilyrose*).

Colophon

The cover photo of a modified Robot Army Delta Bot is by Pawel Szymczykowski. Special thanks to Mark and Sarah for letting Pawel borrow it for the photo shoot.

The cover fonts for *Make: JavaScript Robotics* are URW Typewriter and Guardian Sans. The text font is Adobe Minion Pro; the heading font is Adobe Myriad Condensed; and the code font is Dalton Maag's Ubuntu Mono.